Advances in Intelligent Systems and Computing

Volume 218

Series Editor

J. Kacprzyk, Warsaw, Poland

For further volumes:
http://www.springer.com/series/11156

Pierpaolo Vittorini · Rosella Gennari · Ivana Marenzi
Tania Di Mascio · Fernando De la Prieta
Editors

2nd International Workshop on Evidence-Based Technology Enhanced Learning

 Springer

Editors

Pierpaolo Vittorini
Department of Internal Medicine and
   Public Health
University of L'Aquila
L'Aquila
Italy

Tania Di Mascio
Department of Engineering Computer
   Science and Mathematics
University of L'Aquila
L'Aquila
Italy

Rosella Gennari
Computer Science Faculty
Free University of Bozen-Bolzano
Bolzano
Italy

Fernando De la Prieta
Department of Computing Science
University of Salamanca
Salamanca
Spain

Ivana Marenzi
L3S Research Center
Leibniz University of Hannover
Hannover
Germany

ISSN 2194-5357   ISSN 2194-5365   (electronic)
ISBN 978-3-319-00553-9   ISBN 978-3-319-00554-6   (eBook)
DOI 10.1007/978-3-319-00554-6
Springer Cham Heidelberg New York Dordrecht London

Library of Congress Control Number: 2013937322

Printed on acid-free paper

Springer is part of Springer Science+Business Media (www.springer.com)

Preface

Research on Technology Enhanced Learning (TEL) investigates how information and communication technologies can be designed in order to support pedagogical activities. The Evidence Based Design (EBD) of a system bases its decisions on empirical evidence and effectiveness. The evidence-based TEL workshop (ebTEL) brings together TEL and EBD. The workshop proceedings collects contributions concerning evidence based TEL systems, like their design following EBD principles as well as studies or best practices that educators or education stakeholders used to diagnose or improve their students' learning skills, including students with specific difficulties (e.g. poor/slow readers, students living in impoverished communities or families).

The ebTEL international workshop series was launched under the collaborative frame provided by the European TERENCE project (www.terenceproject.eu). The TERENCE project, n. 257410, is funded by the European Commission through the Seventh Framework Programme for Research and Technological Development, Strategic Objective ICT-2009.4.2, Technology-enhanced learning. TERENCE is building an AI-based Adaptive Learning System (ALS) for reasoning about stories, in Italian and in English, through reading comprehension interventions in the form of smart games. The project also is also developing innovative usability and evaluation guidelines for its users. The guidelines and the ALS result from a cross-disciplinary effort of European experts in diverse and complementary fields (art and design, computers, engineering, linguistics and medicine), and with the constant involvement of end-users (persons with impaired hearing and their educators) from schools in Great Britain and Italy.

The first edition of ebTEL collected contributions in the area of TEL from computer science, artificial intelligence, evidence-based medicine, educational psychology and pedagogy. Like the previous edition, this second edition, ebTEL'13, wants to be a forum in which TEL researchers and practitioners alike can discuss innovative evidence-based ideas, projects, and lessons related to TEL. The best papers of ebTEL'13 were also invited for a special issue of the International Journal of Technology Enhanced Learning (IJTEL) through a dedicated call for papers. The workshop takes place in Salamanca, Spain, on May 22nd–24th 2013.

This volume presents the papers that were accepted for ebTEL 2013. The full program contains 14 selected papers from 6 countries (Finland, Germany, Italy, Spain and United Kingdom). Each paper was reviewed by, at least, two different reviewers, from an international committee composed of 29 members of 9 countries. The quality of papers was on average good, with an acceptance rate of approximately 80%.

We would like to thank all the contributing authors, the reviewers, the sponsors (IEEE Systems Man and Cybernetics Society Spain, AEPIA Asociación Española para la Inteligencia Artificial, APPIA Associação Portuguesa Para a Inteligência Artificial, CNRS Centre national de la recherché scientifique and STELLAR), as well as the members of the Program Committee, of the Organising Committee and of the TERENCE consortium for their hard and highly valuable work. The work of all such people contributed to the success of the ebTEL'13 workshop.

We would like to especially acknowledge the contributions of a TERENCE colleague that, sadly, did not live long enough for seeing the results of our joint work: Emanuele Pianta, whose dedication to work, patient guidance and respect for others' views allowed us to hold our ground in difficult times and grow, all together, in the frame of ebTEL. This volume is for you, Emanuele. *Grazie.*

The Editors
Pierpaolo Vittorini
Rosella Gennari
Ivana Marenzi
Tania Di Mascio
Fernando De la Prieta

Organisation

Steering Committee

Juan M. Corchado	University of Salamanca (Spain)
Fernando De la Prieta	University of Salamanca (Spain)
Rosella Gennari	Free University of Bozen-Bolzano (Italy)
Ivana Marenzi	L3S Research Center (Germany)
Pierpaolo Vittorini	Universy of L'Aquila (Italy)

Program Committee

Mohammad Alrifai	L3S, Leibniz University of Hannover (Germany)
Anthony Baldry	University of Messina (Italy)
Vincenza Cofini	University of L'Aquila (Italy)
Ulrike Cress	University of Tübingen (Germany)
Giovanni De Gasperis	University of l'Aquila (Italy)
Juan Francisco De Paz	University of Salamanca (Spain)
Dina Di Giacomo	University of l'Aquila (Italy)
Tania Di Mascio	University of l'Aquila (Italy)
Gabriella Dodero	Free University of Bozen-Bolzano (Italy)
Peter Dolog	Aalborg University (Denmark)
Erik Duval	Katholike Universiteit Leuven (Belgium)
Marco Fisichella	L3S, Leibniz University of Hannover (Germany)
Wild Fridolin	Open University of the UK (UK)
Ana Belén Gil	University of Salamanca (Spain)
Oscar Gil	University of Salamanca (Spain)
Carlo Giovannella	University of Roma, Tor Vergata (Italy)
Eelco Herder	L3S, Leibniz University of Hannover (Germany)

Ralf Klamma	RWTH Aachen University (Germany)
Ralf Krestel	University of California (Irvine - USA)
Katherine Maillet	Telecom Ecole de Management (France)
Wolfgang Nejdl	L3S, Leibniz University of Hannover (Germany)
Stefano Necozione	University of l'Aquila (Italy)
Elvira Popescu	University of Craiova (Romania)
Sara Rodríguez	University of Salamanca (Spain)
Mario Rotta	Università degli Studi di Firenze (Italy)
Maria Grazia Sindoni	University of Messina (Italy)
Marcus Specht	Open University of the Netherlands (Netherlands)
Sara Tonelli	FBK-irst (Italy)
Carolina Zato	University of Salamanca (Spain)

Local Organising Committee

Juan M. Corchado	University of Salamanca (Spain)
Javier Bajo	Pontifical University of Salamanca (Spain)
Juan F. De Paz	University of Salamanca (Spain)
Sara Rodríguez	University of Salamanca (Spain)
Dante I. Tapia	University of Salamanca (Spain)
Fernando De la Prieta	University of Salamanca (Spain)
Davinia Carolina Zato Domínguez	University of Salamanca (Spain)
Gabriel Villarrubia González	University of Salamanca (Spain)
Alejandro Sánchez Yuste	University of Salamanca (Spain)
Antonio Juan Sánchez Martín	University of Salamanca (Spain)
Cristian I. Pinzón	University of Salamanca (Spain)
Rosa Cano	University of Salamanca (Spain)
Emilio S. Corchado	University of Salamanca (Spain)
Eugenio Aguirre	University of Granada (Spain)
Manuel P. Rubio	University of Salamanca (Spain)
Belén Pérez Lancho	University of Salamanca (Spain)
Angélica González Arrieta	University of Salamanca (Spain)
Vivian F. López	University of Salamanca (Spain)
Ana de Luís	University of Salamanca (Spain)
Ana B. Gil	University of Salamanca (Spain)
Mª Dolores Muñoz Vicente	University of Salamanca (Spain)
Jesús García Herrero	University Carlos III of Madrid (Spain)

Contents

The 1st Release of the TERENCE Learner GUI: The User-Based Usability Evaluation

Maria Rosita Cecilia, Tania Di Mascio, and Alessandra Melonio

Abstract. This paper reports the user-based usability evaluations performed in Italy of the first release of the learner Graphical User Interface (GUI) of the TERENCE project. This project aims at developing an adaptive learning system for training the reasoning about stories' events of the TERENCE learners in Italy and in UK. Learners are 7-11 year old children, hearing and deaf, that have difficulties in correlating the events of a story, making inferences about them, and detecting inconsistencies. The evaluation of the first release of the TERENCE adaptive learning system software prototypes tackles their usability in order to quickly reveal possible usability problems, as well as to address the TERENCE team to solve them, before the large scale evaluation. Moreover, authors try to carried out important general issues related to the experiment performance.

1   Introduction

The main reason to concentrate our effort on evaluating the usability of the TERENCE Graphical User Interfaces (GUIs) before the large scale evaluation mainly derives from the fact that, as well described in the [3] survey, "...the approaches used to evaluate Adaptive Learning Systems (ASLs) are similar in one aspect: they

Maria Rosita Cecilia
Dep. of Life, Health and Environmental Sciences,
University of L'Aquila, 67100, L'Aquila, Italy
e-mail: mariarosita.cecilia@univaq.it

Tania Di Mascio
Dep. of Engineering, Information Science and Mathematics, University of L'Aquila,
67100, L'Aquila, Italy
e-mail: tania.dimascio@univaq.it

Alessandra Melonio
Free University of Bozen-Bolzano, CS Faculty, P.zza Domenicani 3, 39100 Bolzano, Italy
e-mail: alessandra.melonio@inf.unibz.it

P. Vittorini et al. (Eds.): *2nd International Workshop on Evidence-Based TEL*, AISC 218, pp. 1–8.
DOI: 10.1007/978-3-319-00554-6_1   © Springer International Publishing Switzerland 2013

tend to evaluate an ALS as *a whole*, focusing on an *end value* delivered by the system such as the overall user's performance or the user's satisfaction... Evaluating a system as a whole can be acceptable in the field where no acceptable component model of a system can be identified. However, it is not the case for adaptive systems... '.

This paper suggests using a *layered evaluation process*, in which one of the layer is represented by the learning material and an other by GUIs. The authors introduced such an approach to guide designers in the ALS development process. Such a layered approach is in line with the *User-Centred Design* (UCD), used in the TERENCE project, where the evaluation is used iteratively and incrementally to refine the requirements, the design or the development of the system. Moreover, all the evaluation studies reported in [3] stressed the fact that the usability issues of the ALS interfaces have to be solved before starting the evaluation of the ALS in order to minimize bias in the evaluation study of the ALS' usability as "a whole", that is the ALS' pedagogical effectiveness. A thing which is mandatory in UCD as well.

The TERENCE project took up such a two layer up for the learning material and the GUIs, before the large scale evaluation, in two main manners:

- the TERENCE team evaluated the learner material and the GUI prototypes via expert-based evaluations reported in [9],
- the TERENCE team evaluated the refined version of the learner material and the first releases of the GUIs via user-based evaluation, reported in [5].

In this paper, we focus on the user-based evaluation, mainly reported in [5], of the most complex GUI, namely TERENCE learner GUI. The entire learner GUI is available at http://hixwg.univaq.it/learner-gui.html; its design is described in [10].

2 Experiment Description

For the experiments we here describe, we adopted user-based criteria methods like observational evaluation [1], semi-structured interviews [7] and think-aloud protocol [6]. In fact, the approaches used in the literature for evaluating TEL projects are mainly user-based (see [8]). An important reason justifying the usage of user-based approaches in TEL projects is the fact that users are often involved in the design of the projects. Like the other TEL projects, the TERENCE project involves users in the evaluation process. In doing so, the TERENCE team opted for methods that are adequate to the TERENCE main users, that is, 7–11 year old children, and prone to being used in numerous but short inexpensive evaluation sessions. In fact, observational evaluation, semi-structured interviews, and think-aloud protocol are semi-structured methods for examining and reporting problems with the learner GUI in qualitative and quantitative ways.

The reports of the assessments for the learner GUIs usability evaluations and the learning material in [5] is divided as follows: (1) *goals* of the assessment; (2) *participants*, that is, the description of the involved users; (3) *tasks and material*, that is, the description of tasks and material proposed for the experiments; (4) *results*, that is, the description of significant results. This choice is the same we use to structure this section.

2.1   Experiment Goals

The overall goal was to examine whether the sequence of tasks in Table 1 and, more in general, the navigation of the learner GUI were usable for the intended age range. In particular, we also tried investigating the user experience with the learner GUI, more precisely, with:

1. the avatars, and its role in the learner GUI,
2. the stories, whether appealing or not for the learners,
3. the cards of characters, whether interesting or not,
4. the smart games, whether playful or too difficult,
5. the relaxing games, whether sufficiently appealing.

The focus was on identifying areas whether and which improvements should be made prior to the large scale evaluation.

In Italy, it was possible to run several sessions and, by incrementally and iteratively improving on the learner GUI prototype, it was eventually possible to gather also quantitative data where sufficient technical facilities, like a stable wireless, were available.

Table 1 Usability evaluation tasks

Task order	Task description
1	accessing the system via the login page
2	choosing an avatar
3	choosing a book
4	choosing a story in the spatial map of the book
5	browsing and reading the cards of characters
6	browsing and reading a story
7	browsing and playing with smart games
8	browsing and playing with relaxing games

2.2   Experiment Participants

The evaluation in Italy counted 57 learner participants, out of which 16 are deaf, all aged 7–12, and from different locations from the North and the Centre of Italy:

- Centre of Italy: the summer school of the National Laboratories of Gran Sasso (LNGS), nearby l'Aquila; the summer school of Sacro Cuore in Avezzano;

private lodging in Avezzano; a summer school for deaf children in Ciampino, nearby Rome;

- Hearing 12 (7-9 years old); 18 (9-11 years old);
- Deaf 1 (7-9 years old); 8 (9-11 years old).

• North of Italy: the Akademia summer school in Bolzano; the unity of audiology and phonology of the Ca' Foncello Hospital in Treviso.

- Hearing 3 (7-9 years old); 3 (9-11 years old);
- Deaf 5 (7-9 years old); 7 (9-11 years old).

All the participants had used at least once a PC with mouse. In the North of Italy, 5% of the children did not know the tablet and they had never used it. In each session, there were 1 or 2 children per experimenter. Among the experimenters, there were always an expert facilitator.

2.3   Experiment Tasks and Material

At the start of every session, each learner or their educators were asked some questions in order to know the learner's school class and age, and then the experimenter inserted the appropriate login information on behalf of the child. The facilitator informed all children about the goal of the evaluation, that is, to present them a system for helping their peers to better understand a story. The facilitator then asked the children to talk aloud their opinions: since the system was in its infancy, it was important that the children would tell us their opinions while using the system, on what was clear and what unclear, so as to help us improve the system with their valuable feedback.

At this point, the evaluation session started. Every child could interact with the learner GUI by using a 10" tablet or PC with mouse. During the session, the child could perform the sequence of tasks in Table 1.

In particular,

• younger children, whose age is 7–9, read either "La Vacanza Comincia" (The Holiday Begins) or "A Caccia di Delfini" (Dolphin Spotting),
• older children read either "La Mania della Competizione di Benedetto" (Ben's Racing Problem) or "Sofia e il Nano dell'Isola" (Sophia and the Island Dwarf),

and played with the associated smart games that were, on average, 1 per game level:

• 1 who game and 1 what game;
• 4 games that required reasoning about time: 1 before-after, 1 before-while, 1 after-while, 1 before-while-after;
• 2 games that required reasoning about causality: 1 cause or 1 effect, 1 cause-effect.

The learning material is reported and described in [2] and [4]; it is also available at *www.terenceproject.eu/demos*.

During the session, the appointed experimenter was observing and intervening only at critical points, when the learner definitely needed assistance. Qualitative data were thus gathered:

- via direct observations, e.g., of facial expressions, and by tracking comments per tasks;
- via indirect questions to children at critical points, e.g., if the child asks for help, if the child looks lost, if the child looks frustrated.

In the North of Italy, technical facilities, like a stable wireless, allowed for collecting reliable quantitative data through logs. The quantitative data we gathered were:

- session time, that is, the time span in between the start and the end of a session;
- for the reading task, the start and end time for reading the book selected, and the reading time per page;
- for each game instance, the number of correctly resolved game instances, and the time for their resolution before the game was over.

At the end each session, the experimenters run a debriefing phase and a short interview with indirect questions, reported in [5]. The questions were related to the usability and the experience of the learner with the system, i.e., their previous acquaintance with tablets and PCs, as well as whether they thought the story text or illustrations were appropriate for younger/older children, whether they thought the games and the interaction gestures implemented are appropriate for younger/older children, whether the avatar are nice for younger/older children, and what they would like to improve in the GUI.

2.4 Experiment Results

All the usability issues tracked, during the session or the debriefing phase, are reported in details in [5] in a specific category, per country, and should be considered for improving the design of the learner GUI. In general, a category corresponds to a task, e.g., playing with a smart game. Two categories do not correspond to specific tasks, namely, the avatar and the navigation category. Therein, we gathered issues re-occurring in different tasks, and then removed the issues from these tasks. An example is the position of the avatar during the browsing of books and stories, as well as during the reading and playing activities: positioning the avatar in the top-right corner of the screen hides the avatar, and its role in the learner GUI is too passive or unclear. The category named "playing with smart games" is subdivided into subcategories of correlated tasks or issues, e.g., choices available to the learner, so as to facilitate the interventions of those working on the design and requirements. Table 2 briefly reports the usability results divided in positive and negative issues, where

- negative (NEG) issues if they pinpoint specific usability problems,
- and positive (POS) issues if they support design choices or purport a positive user experience.

Table 2 Usability results, divided into negative and positive

Category	Negative	Positive	Number negative	Number positive
Login	lack of human presence or animation	most beautiful page for many children	3.00	2.00
Navigation buttons	next button overlooked next button area too narrow	enjoyable and usable dashboard effect	2.00	1.00
Avatars	no browsing besides the 2nd avatar no clear role of avatar after being chosen no perception of the avatar's growth or the avatar's relations with points no black avatars	nice images male learners choose the male avatars, female learners choose the female avatars	4.00	2.00
Choosing books	book titles too small	children liked the general layout	1.00	1.00
Choosing stories	too big or too poorly coloured padloacks	beautiful spatial map for choosing stories of a book clear spatial map for choosing stories of a book	1.00	2.00
Reading stories	not sufficiently visible page number complaints of too small fonts or not nice font type vocabulary at times too difficult vocabulary or length of sentences too difficult for deaf children in the Centre of Italy illustrations: some children complained about incoherencies between story texts and illustrations or badly resized images; some older children judged illustrations good for younger children; many deaf children complained about lack of vivid colours, and the characters being always the same or the illustrations not being realistic	deaf children in Treviso first read then looked at images for fixing in mind what they had read story plots were generally judged funny and creative, instructive and with a deep meaning children, in particular younger ones, liked the illustration style	5.00	3.00
Interaction with character cards	problems with captions, e.g., too small font size or typos card under focus was not readable information was not read too numerous or confusing, better per story	older children seem more interested in reading the cards than younger children	4.00	1.00
Playing with relaxing games	decontextualised frustrating due to too fast timing or gesture usability issues slicing the fruit reminded of the Fruit Ninja games and, on PCs, children wanted to repeat the same gesture in the toolbox games, children were expecting to be able to rotate images	children generally appreciated the types of games	4.00	1.00
Playing with smart games			14.00	9.00
	Making a choice: not sufficiently usable captions, e.g., too small font size more correct choices in what games time and causality: allow to revise a choice gray-out effects for unavailable choices not always working		4.00	0.00
	Gesture and interaction modes: drag and drop slightly more difficult with mouse, but technical problems make it hard to be usable on touch-screen tablets low affordance for causality games		2.00	0.00
	Feedback: too fast timing too prominent or not well placed yes/no feedback visual metaphor of explanatory feedback for time games was not sufficiently clear for younger children several children are willing to re-read a story		4.00	0.00
	Points and instructions: instructions not read or not sufficiently clear points not noticed or not sufficiently clear		2.00	0.00
	User difficulties: some children complained about the plausibility of solutions of 3 games (1 before-while, 1 cause, 1 cause-effect)	time and causality games are more difficult than the other games time and causality games are more difficult for younger learners than for older learners who and what are entry level games	1.00	3.00
	Other user satisfaction issues: one illustration not matching with the caption	general enjoyment of smart games, e.g., "better than the traditional boring education games" liked what games' and causality games' visual metaphors or animation effects younger children wowed what games the majority of children from the North voted the causality games as the most beautiful drag and drop were considered more challenging and appealing by the majority of learners, and older learners judged it better for them children liked the visual metaphors of the feedback	1.00	6.00
TOTALE			38.00	21.00

3 Discussion and Conclusion

In light of the user-based usability evaluations, we find the following results:

- the interaction with cards needs to be improved; children did not often read the information of the cards, they often quickly looked at the images of the cards, and flickered through these; some children suggested the GUI display only the characters of each story, and not of the entire book;
- children in Italy generally liked the types of relaxing games; however, all children were frustrated when unable to play the games, due to the too fast time-out or the usability of gestures; they were also puzzled by the fact that the relaxing games were not contextualised in the learner GUI (e.g., missing points);
- playing with smart games and reading a story are tasks with the highest number of usability results, which are uniformly distributed between negative and positive; thus, they are likely to be determinant for the success of the software.

Moreover, analysing the overall results described in [5] we find that the causality games are more difficult than the time games, which may be well due to the low affordance of the software version of the causality games, as highlighted by the usability evaluations. It is remarkable that quantitative results in Italy show a correlation between age and resolution of time and causality games; in particular,

- the usability evaluations in Italy show that younger learners had more difficulties with the time and causality games than with the other games, and their highest resolution time was for the time games;
- again the usability evaluations in Italy show that deaf learners had more difficulties with the time and causality games than with the other games, and their highest resolution time was for the time games.

Such results purport that the resolution of time and causality games can give reliable indications to the adaptive engine and, more precisely, whether this can move the learners from one story level to another, as designed in [2].

Another issue, is the need of presenting instructions in a different format, clearer and more appealing. In particular, during the usability evaluations children tended not to read the instructions or these were not sufficiently clear. Moreover, some deaf children needed further assistance the first time they played with a new game level. The usage of contextualised tutorials the first time the learners play with games may be beneficial.

According to the usability evaluations, some of the vocabulary seems at points too difficult, also at the simplest story level. However, it was found that, when prompted to infer the meaning of unknown words from the text, the deaf children generally were able to do it. All deaf children were invited to re-read the story in order to allow them to perform the games.

In conclusion all learners, hearing and deaf, were able to perform the designed tasks. Both of them preferred playing instead to reading. All of them do not really understand the role of the avatar though all learners like avatar. This is due of the fact that avatars were not very well contextualised. The difference between the two types of users is tangible in the administration of the tasks: the presence of the LIS Italian translator makes experiments more low and more time consuming. Conversely, deaf children are more critics and they are more interested to reveal us the issues to check and correct.

Acknowledgements. This work was supported by the TERENCE project, funded by the EC through FP7 for RTD, ICT-2009.4.2.

References

1. Paramythis, A., Weibelzahl, S., Masthoff, J.: Layered evaluation of interactive adaptive systems: framework and formative methods. User Modeling and User-Adapted Interaction (2010)
2. Arfé, B.: Repository of Stories, Deliverable D2.2. Tech. rep., TERENCE project (2012)
3. Brusilovsky, P., Karagiannidis, C., Sampson, D.: Layered evaluation of adaptive learning systems. In: Int. J. Cont. Engineering Education and Lifelong Learning, vol. 14(4-5) (2004)
4. Gennari, R.: Repository of Textual Smart Games, Deliverable D4.3. TERENCE project (2012)
5. Mascio, T.D., Gennari, R., Vittorini, P.: Small scale evaluation: design and results. TERENCE project (2012)
6. Olmsted, E., Murphy, E., Hawala, S., Ashenfelter, K.: Think-aloud protocols: A comparison of three think-aloud protocols for use in testing data-dissemination web sites for usability. In: Proc. of CHI 2010: 1001 Users (2010)
7. Russo, J.E., Johnson, E.J., Stephen, D.L.: The validity of verbal protocols. Memory and Cognition 17(6), 759–769 (1989)
8. Slegers, K., Gennari, R.: State of the Art of Methods for the User Analysis and Description of Context of Use. Tech. Rep. D1.1, TERENCE project (2011)
9. Mascio, T.D., Gennari, R., Vittorini, P.: Expert- based evaluation. TERENCE project (2012)
10. Vittorini, P.: Integrated system development, first release. TERENCE project (2012)

Promoting Digital Skills and Critical Awareness through Online Search and Personal Knowledge Management: A Case Study

Maria Cinque and Maria Bortoluzzi

Abstract. In the knowledge society the processes of learning and knowledge management take place, very often, online and in social online environments, thus creating issues of complexity and sustainability related to cognitive processes of learning that students - even at university level - are not always able to recognize and cope with. In this paper we present a research case study carried out at the University of Udine with a group of first year students of Multimedia Communication and Technology during the course of English language (*Englishes and Media Communication in a World Context*). The aim of the research was to determine whether specific activities can enhance the development of skills for lifelong learning, such as the ability to search the Internet and use online resources to promote continuous education and learning to learn. Quantitative and qualitative data were gathered within a framework of Personal Knowledge Management and the results are guidelines potentially useful both for teachers and learners.

Keywords: PKM (Personal Knowledge Management), Research and information management using online media and tools, Management of online presence, Effective teaching techniques and strategies for learning.

1 Introduction

In contemporary networked society we make an intense and pervasive use of the network applications in the management of our personal knowledge. While using the available electronic systems, we are gaining awareness about the importance of acquiring more critical, creative and ethical skills in the knowledge management

Maria Cinque
Università degli Studi di Udine / Fondazione Rui
e-mail: m.cinque@fondazionerui.it

Maria Bortoluzzi
Università degli Studi di Udine
e-mail: maria.bortoluzzi@uniud.it

P. Vittorini et al. (Eds.): *2nd International Workshop on Evidence-Based TEL*, AISC 218, pp. 9–16.
DOI: 10.1007/978-3-319-00554-6_2   © Springer International Publishing Switzerland 2013

processes. Issues connected with identity management and protection, reputation management, information overload, 'fair' and effective use of network resources, are becoming crucial.

In order to promote "effective uses" of web resources, we need to reflect on the way technologies are used in education. In this paper we present a research case study carried out at the University of Udine with a group of first year students of Multimedia Communication following the course of English as a foreign language (course title: *Englishes and Media Communication in a World Context*). The main aim of this case study is to gather information on the possibility of enhancing – through searching activities on the Internet - the skills of the students in terms of learning strategies, motivational variables and learning outcomes. After the theoretical background (Section 2), Section 3 illustrates the project development and in Section 4 some results are analysed. A four-step process for Evidence Based Practice is provided in the conclusions (Section 5).

2 Background

As the "digital natives" debate has demonstrated, there is a discrepancy between myth and reality of students' seemingly widespread use of technologies [1] and there are also clear disparities between the education rhetoric and educational realities of social media use [2]. Many studies [3, 4] have pointed out that digital natives are not necessarily digitally competent.

Digital competence is recognized as one of the eight key competences for lifelong learning by the European Union. As [5] points out there are many definitions of digital competence and there are many overlapping concepts, such as *digital literacy, ICT literacy, media literacy* etc. Digital literacy is the broadest concept: it includes the main aspects of the two other fields but also responsible and effective use of digital tools for personal tasks while benefiting from people's networks.

Digital competence is a transversal key competence that enables acquiring other key competences (e.g. language, mathematics, learning to learn, cultural awareness). It is related to many of the so-called 21st century skills which should be acquired by all citizens, in order to ensure their active participation in society.

There are many different versions of the 21st century skills framework. Ananiadou and Claro's work [6] elaborated for OECD (Organisation for Economic Co-operation and Development) focuses on three dimensions: information, communication, ethics and social impact. We used this framework to analyse digital tasks produced by students during our case study.

Our work also refers to another wide area of study, PKM (Personal knowledge management), which is defined as a collection of processes that a person uses to gather, classify, store, search, retrieve, and share knowledge in his/her daily activities [7] and the way in which these processes support work activities. It is a response to the idea that knowledge professionals increasingly need to be responsible for their own growth and learning [8].

The case study we present is based on a PKM framework [9] in which such competences are divided into two main groups: Basic and Higher Order PKM skills. The former include three macro-competence categories: create, organise and share. The Higher Order skills and competences are grouped into four main categories: connectedness, ability to balance formal and informal contexts, critical ability and creativity.

3 Research and Information Management

3.1 Goals, Contents and Expected Learning Outcomes

The aim of the research was to determine whether specific activities can enhance the development of skills for lifelong learning, such as the ability to search the Internet and use online resources to promote continuous education and improve on learning to learn. The main idea was to provide an overview of useful web tools for learning, a series of practical references and a 'toolbox' to be used in different contexts (at university, but also self-learning and professional training).

Our general goals were: 1. to analyze different ways of dealing effectively with the dynamics of the online tools and facilities in order to promote Higher Order PKM skills; 2. to identify tools supportive of such skills or conducive to learning and improving on them; 3. to provide suggestions for the actual application of such tools in the daily practice of teachers.

A brief course on Research and Information Management was designed and implemented for a group of first year students of Multimedia Communication and Technology at the University of Udine. The seminar was carried out within the course of English as foreign language (competence level from B1 to C1 of the Common European Framework of Reference) with over 100 participants[1]. The course was *blended*: students were in class with the English teacher (M. Bortoluzzi); brief synchronous seminars were carried out online via Skype and social media (M. Cinque). Students were asked to perform assigned tasks both at university (using the wi-fi network) and at home.

Initially we administered a questionnaire to investigate on student skills and attitude towards technology. The areas of investigation included: metacognition, motivation and use of ICT tools for learning.

A wiki was used as a 'hub' for all the activities and the contents: http://pordenonestm.pbworks.com; many other different tools (Google and other search engines, YouTube, iTunes, Delicious, Wordle, software for mind mapping etc.) were used for course-work tasks.

The course – carried out between February and March 2012 – lasted 20 contact hours[2], including brief seminars and activities on the following subjects: Search strategies on the Internet (29/02/12); webquests and enquiry based learning (01/03/12); Google services and tools and criteria to evaluate the credibility of web sites and resources (07/03/12); top tools for learning (08/03/12); PLE, Personal Learning Environment (14/03/12).

At the end of the course, students were expected to be able to: 1. search Google effectively and precisely, by custom-izing it; 2. know when to use other search

---

[1] For our research we gathered data from and about 64 students.
[2] Plus a 4-hour session for the final presentations.

engines and web di-rectories; 3. evaluate what they find on the web; 4. create their own PLE; 5. make an effective use of resources found on the web.

Different tools of assessment were used: 1. user questionnaires (the survey on metacognition, motivation and ICT which was administered twice: before and after the course; intermediate and final tests on acquired knowledge); 2. analysis of digital tasks (coursework tasks and final project work); 3. secondary data gathering and analysis (feedback on the course; interview with the teacher).

3.2 Digital Tasks

During the course different tasks were assigned to be performed either during the lessons (in the wifi area of the university) or at home (coursework tasks).

During the lessons students were asked to brainstorm and choose efficient, suitable search terms; create a tag cloud with the results of the brainstorming; categorize the search terms and create a mind map of the search; answer a short search quiz; analyse the web results and evaluate the credibility of websites. The coursework tasks included the following: create a personalized search engine using Google CSE; search on YouTube creative videos on arts or videoart, tag them, and prepare an oral or digital presentation about them; search on iTunes audio or video podcasts on a specific topic and prepare a presentation; search on Twitter what people are saying about a specific topic, categorize the results (in keywords) and create a tag cloud using Wordle; create a Facebook page or a Facebook Group on a positive social or cultural initiative and add ideas to promote the group or the page; using delicious, collect bookmarks concerning websites on a specific topic, like leadership, and prepare a 2-minute oral presentation on the results.

For the final assessment students were asked to prepare – in group of 2-3 people – a multimedia piece of work (a powerpoint presentation, a short video, photo presentation, website, a wiki, a poster, a chart, a sound file, etc.) choosing among one of these three tasks: *My digital identity*, i.we a presentation on their data on the web and on the way to manage their online reputation; *PLE description and design*, a description of the web technologies used for personal and academic purposes; *Spreading good news. Internet branding for no profit initiatives*, i. e. the application of corporate branding techniques that could help make a small no-profit initiative more popular on the web.

4 Some Results

4.1 Metacognition, Motivation and ICT Use

In [10] the detailed description of data gathering and analysis is presented and discussed; here we summarise some of the main findings. The user questionnaire consisted of three parts: 1. Learning strategies; 2. Motivational variables and learning outcomes; 3. ICT use for learning. Our aim was to analyse the relationships between learning strategies and motivation and the use of ICTs.

Se calculated mean, standard deviation and correlation among the following variables (each one was represented by almost 10 items): *Learning strategies and organisational abilities*, i. e. (SO) Organisation strategies; (SA) Self-assessment strategies; (SE) Elaboration strategies; (SM) Metacognitive sensitivity; (SPP) Strategies to prepare for a test; *Motivational variables*, i. e. (OOA) Learning goal orientation; (OOP) Approach dimension of performance goal orientation; (OOE) Avoidance dimension of performance goal orientation; (A) Self-efficacy; *Learning outcomes*, i. e. (V) assessment; (S) satisfaction.

As shown in Fig. 1, all variables have positive correlations.

	OOA	OOE	OOP	A	SO	SA	SE	SPP	SM	V	S
OOA	1										
OOE	,068	1									
OOP	,437**	,425**	1								
A	,296*	,288*	,300**	1							
SO	,548**	,326*	,433**	,377**	1						
SA	,230**	,326**	,278**	,157	,173	1					
SE	,441**	,381*	,454**	,336**	,632**	,286**	1				
SPP	,535**	,368**	,519**	,386**	,621**	,383**	,533**	1			
SM	,520**	,274*	,472**	,352**	,614**	,294**	,683**	,665**	1		
V	,542**	,363**	,433**	,556**	,514**	,206**	,584**	,464**	,542**	1	
S	,480**	,105	,363**	,112	,446**	,112	,637**	,415**	,530**	,450**	1

** The correlation is significant at level 0.01 (2-code)
* The correlation is significant at level 0.05 (2-code)

Fig. 1 User questionnaires: correlation between metacognition, motivational variables and learning outcomes

For the third part of the questionnaire, following a recent study approach [13] in order to identify the different ICT uses and to reduce the number of variables, we performed a factor analysis of principal components with Varimax rotation on the use of ICTs. This allowed us to establish four factors: factor 1 is related to communication among students (*fora* and chats), surfing the Internet and the online consultation of newspapers and magazines (*Social Use*); factor 2 comprises four elements related to the use of professional tools such as databases, web page design, etc. (*Technical Use*); factor 3 is composed of indicators related to office software use such word processing software, spreadsheet software, presentation software, etc. (*Academic Use*); factor 4 comprises e-learning Software Platforms and e-mail (*Educational Platform Use*).

4.2 Analysis of Digital Tasks

The issues connected to the project works performed during the course were discussed in classroom and received qualitative feedback. We assessed students' ability of finding required information effectively and efficiently, using information effectively to accomplish a specific purpose, accessing and using information ethically and legally. We also assessed their critical awareness towards the topics, namely how information found on the Internet was evaluated and used; how information found on the Internet was organised, managed and presented.

For the final project works, following the OECD framework [6], we created indicators based on the three dimensions: information, communication, ethical and social impact.

Table 1 Indicators for the evaluation of digital tasks (created on the base of [6])

Information	Information as a source	Searching
		Evaluating
		Organising information
	Information as a product	Restructuring & modelling information
		Development of own ideas (knowledge)
Communication	Effective communication	Sharing & conveying the results or outputs of information
	Collaboration & Virtual interaction	Reflection on others' work
		Creation of communities
Ethics and social impact	Social responsibility	Applying criteria for a responsible use at personal and social levels
	Social impact	Development of awareness about the challenges in the digital age

Task 1 - *My digital identity* - was chosen by 9 groups that produced different kind of multimedia work (1 video, 1 Prezi presentation, 7 Power Point presentations); Task 3 - *Spreading good news': Internet branding for no profit initiatives* – was chosen by 11 groups (1 website, 1 video, 9 presentations; 7 works concerned already existing initiatives, 4 works concerned new initiatives); Task 2 – *PLE description and design* – was chosen by 1 group.

Student tasks were mapped onto the three dimensions that highlighted different kinds of 'behaviour' [10]. Some groups limited the search to authoritative sources in order to find useful and relevant information. Other groups were able to edit and share in original ways the information they had found using ICT tools. The communicative aspect was particularly relevant, not only because the main course was in English and about English for media communication (English was a foreign language for everybody, students and teachers), but also in relation to the developing skills to "process, transform and re-process the information". Although we could not monitor the process and dynamics of group work, from the results we could find clear evidence of the type of collaboration (and interaction) that the members of each group had implemented online and offline.

Finally, taking into consideration students' work in terms of contents and presentation skills, it became evident whether the students had or had not set for themselves the problem of a responsible use of the digital resources – at a personal and social level – recognizing the potential risks, and the need for rules that could promote appropriate social interaction on the web. In particular, those groups that had created branding initiatives for non-profit organisations demonstrated the ability to appreciate the social, economic and cultural rights for individuals and the society that the network can offer and took in considerations the challenges and opportunities of digital affordances.

5 Conclusions

From the findings of this case study we can draw a complex picture of PKM in which individual instances (the development of personal skills) converge with the technological aspects and the social dimension of the digital resources. PKM skills express 'learning behaviours' such as problem solving, interaction with teacher and other learners, self-correction, critical reflection, competence improvement, meaning making, experiential learning. The process of learning PKM skills follows the four step of EBP (Evidence-based Practice) [12]: formulating an answerable question; information search; reviewing of information and critical appraisal; employ the results in one's practice/work.

Futhermore learning PKM skills depends on: 1. personal and environmental conditions which promote learning; 2. teaching decisions the educator and the learner need to make to engage successfully in learning; 3. the characteristic behaviours of effective adult learning. The link between these three elements is new and evidence-based, both in design and practice. This overview provides a guide for teachers wishing to adopt the adult learning approach while using technology for their courses.

The gathered evidence suggests that it would be necessary to carry out further research aimed at analysing the relationships among the basic conditions that influence the PLE (Personal learning environment), including both personal learning factors and environmental conditions, and PLN (Personal learning network) that represents the people a learner interacts with and derives knowledge from.

References

1. Selwyn, N.: The digital native – myth and reality. Aslib Proceedings 61(4), 364–379 (2009)
2. Selwyn, N.: Social media in higher education. In: The Europe World of Learning 2012, 62th edn. Routledge, London (2011),
   http://www.educationarena.com/pdf/sample/
   sample-essay-selwyn.pdf (verified on January 10, 2013)
3. Bennett, S., Maton, K., Kervin, L.: The 'digital natives' debate: A critical review of the evidence. British Journal of Educational Technology 39(5), 775–786 (2008)
4. Li, Y., Ranieri, M.: Are 'digital natives' really digitally competent?—A study on Chinese teenagers. British Journal of Educational Technology 41(6), 1029–1042 (2010)

5. Ala-Mutka, K.: Mapping Digital Competence: Towards a Conceptual Under-standing. Sevilla: Institute for Prospective Technological Studies (2011), http://ftp.jrc.es/EURdoc/JRC67075_TN.pdf (verified on January 10, 2013)

6. Ananiadou, K., Claro, M.: 21st Century Skills and Competences for New Millennium Learners in OECD Countries. OECD Education Working Papers, 41 (2009)

7. Grundspenkis, J.: Agent based approach for organization and personal knowledge modelling: knowledge management perspective. Journal of Intelligent Manufacturing 18(4), 451–457 (2007)

8. Smedley, J.: Modelling personal knowledge management. OR Insight 22(4), 221–233 (2009)

9. Cigognini, E.: PKM – Personal Knowledge Management: cosa vuol dire essere una persona istruita nel XXI secolo? Formare 66 (2010), http://formare.erickson.it/wordpress/it/2010/ pkm-personal-knowledge-management-cosa-vuol-dire-essere- una-persona-istruita-nel-xxi-secolo/ (verified on January 10, 2013)

10. Cinque, M.: Reti di apprendimento e gestione autoregolata della conoscenza. Utilizzo di modelli di social computing per l'acquisizione di competenze metacognitive e creative nel lifelong learning. PhD Thesis – University of Udine (2013)

11. Valentín, A., Mateos, P.M., González-Tablas, M.M., Pérez, L., López, E., García, I.: Motivation and learning strategies in the use of ICTs among university students. Computers & Education 61, 52–58 (2013)

12. Nordenstrom, J.: L'EBM sulle orme di Sherlock Holmes. Il Pensiero Scientifico Editore, Roma (2008)

The TERENCE Smart Games Revision Guidelines and Software Tool

Vincenza Cofini, Tania Di Mascio, Rosella Gennari, and Pierpaolo Vittorini

Abstract. TERENCE is an FP7 ICT European project, highly multidisciplinary, that is developing an adaptive learning system for supporting poor comprehenders and their educators. The paper introduces the automatic smart games generation process in TERENCE and presents the guidelines for the manual revision as well as the software system supporting it.

Keywords: Smart games, manual revision, guidelines.

1 Introduction

TERENCE [1] is an FP7 ICT European project, highly multidisciplinary, that is developing an adaptive learning system for supporting poor comprehenders and their educators. Its learning material are stories and games. The games are specialised into relaxing games, which stimulate visual perception and not story comprehension, and smart games [2], which in TERENCE stimulate inference-making for story comprehension [3].

Vincenza Cofini · Pierpaolo Vittorini
Dep. of Life, Health and Environmental Sciences,
University of L'Aquila, 67100, L'Aquila, Italy
e-mail: vincenza.cofini@cc.univaq.it,
        pierpaolo.vittorini@univaq.it

Tania Di Mascio
Dep. of Engineering, Information Science and Mathematics, University of L'Aquila, 67100,
L'Aquila, Italy
e-mail: tania.dimascio@univaq.it

Rosella Gennari
Faculty of Computer Science, Free University of Bozen-Bolzano (FUB),
39100, Bolzano, Italy
e-mail: gennari@inf.unibz.it

P. Vittorini et al. (Eds.): *2nd International Workshop on Evidence-Based TEL*, AISC 218, pp. 17–24.
DOI: 10.1007/978-3-319-00554-6_3   © Springer International Publishing Switzerland 2013

The smart games stimulates the comprehension in terms of: the story characters (who games), the events that take place in the story (what games), the temporal ordering of such events (before/after, while/after, before/while, before/while/after games), and the causal relations (cause, effect, cause/effect games) [4, 5]. The adopted game framework is a generalisation of the EMAPPS one [6].

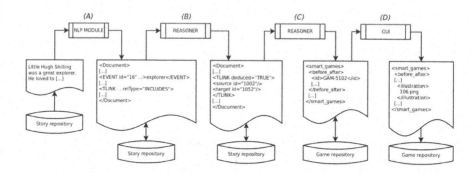

Fig. 1 The smart games generation workflow

Fig. 1 summarises the smart games generation process. The process is made up of the four phases described below:

Phase A. Stories are stored as flat text in a story repository. From a story text, an NLP module generates an annotated story. The annotations follow a variant of the TimeML language [7], that was extended in [8] so to include new tags useful for the TERENCE smart games. For instance, the ENTITY and CLINK tags aim, respectively, at representing the entity related to an event (the EVENT tag is already present in the TimeML standard), and the causal relation between two events. After the annotation process, the annotated story is stored in the same repository;

Phase B. The output of the annotation module is initially checked by a reasoner in terms of consistency of the annotations, i.e., by detecting any eventual temporal inconsistency (e.g., if a relation between two events E_1 and E_2 is that E_1 is before E_2, there cannot exist other relations stating that E_1 is after E_2). If inconsistencies are found, the annotations cannot be stored[1]. Otherwise, the annotations are enriched by adding deduced temporal relations as further TLINK tags [9]. This new consistent and enriched story is also stored in the story repository;

Phase C. Starting from the consistent and enriched story, the reasoner module generates automatically instances of smart games. For instance, for a WHO-game (i.e., who is doing something in the story) related to a certain event, hereafter called *fixed event*:

---

[1] Techniques of relaxation are under development, that will lead to a "self-reparation" process concerning the inconsistent annotations.

- the ENTITY that participates in the fixed event with a role of protagonist is selected as the correct answer;
- other two entities that are not related to the fixed event are added as wrong answers;
- the question that must be asked to the learner is generated through a text-generation module (e.g. if the fixed event is that "Ernesta is riding[2] a bike", the question will be "Who is riding a bike?").

The resulting games are then stored in the game repository;

Phase D. Finally, a manual revision of the generated smart game instances takes place through a dedicated GUI (Sec. 2), by following an ad-hoc written set of guidelines (Sec. 3).

2 GUI for Manual Revision

The GUI is designed with the twofold aim of

1. providing the context (i.e. the story from which the games were generated) to the experts during the revision;
2. keeping a link from the game under revision and the location (in the story text) of both the fixed event and choices.

Accordingly, the GUI is designed as follows: the story text is shown on leftmost area of the screen, and the interface where the revisions actually take place is placed on the right (as of point 1 above). Moreover, when either the fixed event or the choices are focused, the position of the related event/entity is highlighted in the story text (as of point 2 above).

Fig. 2 depicts the current interface structure. Few notes about the part of the interface concerning the actual smart games revision follow. The interface shows the game type on top and enables to change the game question (only for who-games). Right below (in a greyed box), it is possible to change the fixed event (on the right the corresponding word is highlighting, see dotted arrow) as well as its text. Furthermore, on the bottom, the choices are listed. For each choice, the right/wrong events/entities, the related text, and the correctness of the choice can be modified.

To enable such a kind of flexibility, the GUI is developed as a web application written by using the jQuery library[3]. Furthermore, the smart games are taken from a repository organised in terms of books/stories/versions, placed on a server under revision control, and – when finalised – stored into the story repository.

---

[2] The verb "to ride" will be annotated as event.

[3] jQuery is JavaScript Library that simplifies HTML document traversing, event handling, animating, and Ajax interactions for rapid web development.

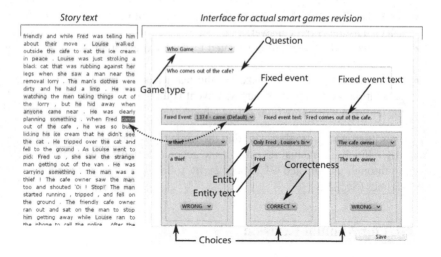

Fig. 2 GUI structure

3   Guidelines for Manual Revision

TERENCE provides stories rewritten into four different levels of increasing difficulty. Level 1 is written for deaf learners, level 2 for skilled deaf learners, level 3
for poor comprehenders, and level 4 for skilled learners [10]. However, the smart
games are all generated from story level 1, so to be comprehensible to all our types
of learners. Accordingly, also the guidelines detailed below are written so that the
resulting smart games shall at least suit the deaf audience (e.g. by preferring short
sentences, possibly without subordinates and coordinates, possibly minimising the
referential expressions especially if complex).

Fig. 3 Smart games selection

The last updated GUI version is available at http://hixwg.univaq.it/TERENCE-expert-gui/ . In order to load the games for a story, the reviewer must act as follows:

1. first, click on "Open", choose a book from the tab menu in your language (EN, IT), then click on VERSION 1 (only) of a story [4] (Fig. 3);
2. then, the "Editor" link actually opens the editor (Fig. 2), that enables the following three main functionalities:

   a. revise existing games;
   b. create new smart games;
   c. delete a game.

   Each is explained in details below.

3.1 Revision

Task 1. Revise the sentences according to the following general guidelines, whenever possible:

1. the sentence must be related to the selected event;
2. prefer short sentences, possibly without subordinates and coordinates, possibly minimising the referential expressions especially if complex (example: "drag the two cards below those over them");
3. add

   a. location, or
   b. a temporal/causal/final subordinate.

   though only when (a) or (b) above are mandatory for identifying unequivocally the event described in the sentence. For example: by given the sentence "Victor approaches Ugolino and Pietro", it should be revised as "Victor approaches Little Hugh and Peter before bringing them the cake" in order to distinguish the event "approaching before bringing the cake" from the event "approaching for bringing the cake" (the two "approaches" take place in different moments of the story);
4. prefer present tense, active form;
5. substitute direct speech with indirect speech. Example: "Giovanni says: "drink the milk!"" should be rewritten as "Giovanni says to drink the milk";
6. report the revision in a spreadsheet file.

Task 2. If the game is wrong (e.g., the solution chosen as BEFORE is wrong), proceed as follows, whenever possible:

1. select through the drop down list, for revision

   a. the entity, e.g., characters, that is right or wrong, in case of who games;
   b. the event, that is right or wrong, in case of the other games.

---

[4] In TERENCE, the stories are available in different levels of difficulties. Version 1 is the easiest, while Version 4 is the hardest to comprehend. The smart games in TERENCE are always generated from the simplest story version.

by always preferring

a. characters to other entities, in case of who games,
b. action verbs than description verbs (better "do, eat, knock, bring" than "say that, tell that, seems that"...) in case of other games;
c. events that have a duration span in the story that is rather clear to the child.

2. revise the sentence possibly in the same way (same meaning) in the different games of the story;
3. report the new choice in a spreadsheet file.

Subtask 2.1. In case of WHO games,

1. revise the CORRECT solution by taking it from the drop-down menu that comes before "Other"; if the correct entity (e.g., Ben) is missing in the menu, use on its behalf another entity (e.g., "child") that does not appear in another WHO game;
2. revise the WRONG solution by taking it from the part that starts with "Other" in the drop-down menu.

Subtask 2.2. In case of all the other smart games, revise a WRONG solution as follows:

1. if possible, keep the same event, otherwise select a WRONG event from the same story that must be different than

a. the correct solution and the other wrong solution in case of WHAT games,
b. the BEFORE, AFTER, WHILE solutions in the time games,
c. the CAUSE, EFFECT solutions in the causality games.

2. rewrite the related sentence by changing the subject/object. For instance, let us suppose that in a WHAT game the expert GUI displays

- the event "bring" with sentence "Victor brings the cake" as WRONG (however, the event actually takes place in the story, therefore it should not be marked as WRONG),
- the event "put" with sentence "Little Hugh put the fork into his mouth" as CORRECT,
- the event "wear" with sentence "Little Hugh wears spectacles" as the other WRONG solution (and this is indeed wrong, as it happens in other story).

In such a case, the reviewer can:

a. select the event "sit" with sentence "Little Hugh and Peter sit at the table" (that does not occur in another WHAT game for the chosen story and is different than the correct solution and the other wrong solution for this game),
b. rewrite the sentence as "Victor sits at the table" (this is an event that does NOT happen in the story).

Similarly we proceed in temporal/causal games for WRONG events so as to make them plausible wrong events and minimise the work of illustrators.

Task 3. Save using the SAVE button.

3.2 Creation

Task 1. In order to create a new game, click on the ADD NEW button.
Task 2. Select the game type.
Task 3. Insert a new game id as follows: *bookX-storyY-version1-new-N*, where X is
the book number, Y is the story number and N is the number for the new game: $N=1$
if it is the first new game, $N=2$ if it is the second new game, etc.
Task 4. Proceed as of "Revision" (Subsec. 3.1).

3.3 Delete

Avoid to delete a game. Try to revise it, if necessary, explain the reasons for deletion
in the note field.

Task 1. Use the DELETE button to delete a game.
Task 2. Create a new game of the same type (e.g., WHAT) as in "Creation" (Subsec.
3.2).

4 Conclusion

The paper summarised the software and methodological basis for the manual revision process that takes place after the automatic generation process. The paper
introduces the rationale behind the guidelines that were used by the staff that took
care of actually revising the smart games, whose results are reported in [11].

Acknowledgements. This work was supported by the TERENCE project, funded by the EC
through FP7 for RTD, ICT-2009.4.2.

Special Thanks to the Developer: A. Paolucci.

References

1. TERENCE Consortium: TERENCE web site
2. Hedberg, S.R.: Executive Insight: Smart Games: Beyond the Deep Blue Horizon. IEEE
   Expert 12(4), 15–18 (1997)
3. Cofini, V., Di Giacomo, D., Di Mascio, T., Vittorini, P.: Evaluation Plan of TERENCE:
   when the User-centred Design Meets the Evidence-based Approach. In: Proc. of the
   ebTEL 2012 Workshop co-located with PAAMS 2012. Springer (2012)
4. Pintado, F.D.L.P., Mascio, T.D., Gennari, R., Marenzi, I., Vittorini, P.: The TERENCE
   Smart Games: Automatic Generation and Supporting Architecture. In: 1st International
   Workshop on Pedagogically-driven Serious Games (September 1, 2012)
5. Cofini, V., de la Prieta, F., Di Mascio, T., Gennari, R., Vittorini, P.: Design Smart Games
   with requirements, generate them with a Click, and revise them with a GUIs. Advances
   in Distributed Computing and Artificial Intelligence Journal 1(3), 59–68 (2012)

6. EMAPPS consortium: EMAPPS Game Framework, `http://emapps.info/eng/Games-Toolkit/Teachers-Toolkit/Games-Creation/Framework-for-Game-Design` (retrieved January 2012)

7. TimeML working group (2011), `http://www.timeml.org/` (retrieved August 2011)

8. Moens, S.: Deliverable 3.1: State of the Art and Design of Novel Annotation Languages and Technologies. Technical Report D3.1, TERENCE project (2012)

9. Gennari, R.: Deliverable 4.1: M.State of the Art and Design of Novel Intelligent Feedback Deliverable 4.1. Technical report, TERENCE project (2011)

10. Arfé, B.: Deliverable 2.2: Repository of Stories. Technical Report D2.2, TERENCE project (2012)

11. Vincenza, C., Gennari, R., Vittorini, P.: The manual revision of the TERENCE italian smart games. In: Vittorini, P., Gennari, R., Marenzi, I., Di Mascio, T., De la Prieta, F. (eds.) 2nd International Workshop on Evidenced-Based TEL. AISC, vol. 218, pp. 25–32. Springer, Heidelberg (2013)

The Manual Revision of the TERENCE Italian Smart Games

Vincenza Cofini, Rosella Gennari, and Pierpaolo Vittorini

Abstract. TERENCE is an FP7 ICT European project, highly multidisciplinary, that is developing an adaptive learning system for supporting poor comprehenders and their educators. The paper describes the automatic smart games generation process in TERENCE, motivates the need for a manual revision and describes it in detail. The paper thus provides a thorough insight in understanding the quality level of the automatic smart games generation process in TERENCE, and the time/effort needed for their manual revisions.

Keywords: Smart games, manual revision, NLP, reasoning.

1 Introduction

TERENCE [10] is an FP7 ICT European project, highly multidisciplinary, that is developing an adaptive learning system for supporting poor comprehenders and their educators. Its learning material are stories and games. The games are specialised into relaxing games, which stimulate visual perception and not story comprehension, and smart games, which stimulate inference-making for story comprehension [8, 1, 2].

In brief, the TERENCE plan for stimulating story comprehension consists of the following increasingly demanding tasks [4]: firstly, it makes the learner reason about

Vincenza Cofini · Pierpaolo Vittorini
Dep. of Life, Health and Environmental Sciences, University of L'Aquila,
67100, L'Aquila, Italy
e-mail: vincenza.cofini@cc.univaq.it,
    pierpaolo.vittorini@univaq.it

Rosella Gennari
Faculty of Computer Science, Free University of Bozen-Bolzano (FUB),
39100, Bolzano, Italy
e-mail: gennari@inf.unibz.it

P. Vittorini et al. (Eds.): *2nd International Workshop on Evidence-Based TEL*, AISC 218, pp. 25–32.
DOI: 10.1007/978-3-319-00554-6_4   © Springer International Publishing Switzerland 2013

the characters that are in the story, then about the events that take place in the story, hence about temporal relations among the events, and finally about causal-temporal relations among events (see taxonomy in Fig. 1). Accordingly, factual, temporal, and causal smart games are the actual implementation of the corresponding comprehension tasks.

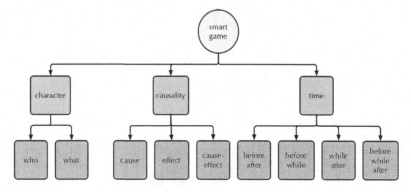

Fig. 1 Smart games taxonomy

In TERENCE, smart games are automatically generated as follows (see Fig. 2):

Phase A.   Firstly, from a story text contained in the story repository, an NLP module generates a story annotated with a variant of the TimeML language, that was extended in [9] with tags that are relevant for the TERENCE smart games. For instance, the ENTITY and CLINK tags aim, respectively, at representing the entity related to an event, and the causal-temporal relations between two events. The annotated story is then stored in the same repository.

Phase B.   Then, a reasoner checks the consistency of the annotations, detects the eventual temporal inconsistencies, and enriches the annotations by adding deduced temporal relations as further TLINK tags [6]. This new consistent and enriched story is also stored in the story repository.

Phase C.   Starting from the consistent and enriched story, the reasoner module generates automatically instances of smart games. For instance, to create a WHO-game related to a certain event [7, 5]:

- the ENTITY that participates in the event with a role of protagonist is selected as the correct answer;
- other two entities that are not related to the event and are different from the entity selected above are added as wrong answers;
- the question asked to the learner is generated through a text-generation module (e.g. if the event is that "Ernesta is riding[1] a bike", the question will be "Who is riding a bike?").

The resulting games are then stored into the game repository.

---

[1] The verb "to ride" is detected as an event.

Phase D.   Finally, a manual revision of the generated smart game instances takes place, where the related visuals (e.g. background illustrations, buttons) are also specified.

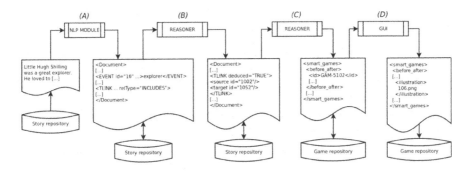

Fig. 2 The smart games generation workflow

It is worth remarking that the errors introduced in the automatic annotation process ("Phase A" mentioned above) influence the quality of all subsequent phases. For instance:

- badly recognised TLINKs (e.g. a BEFORE relation detected as an AFTER relation) lead to the deduction of wrong additional relations. Consequently, the smart games that includes such wrong relations may have temporal games with wrong solutions marked as correct;
- poor annotations may:
  - prevent the generation of some classes of games. For instance, without TLINKs with relation type of INCLUDES/IS_INCLUDED/OVERLAPS, the procedure will not be able to generate any BEFORE/WHILE, WHILE/AFTER, BEFORE/WHILE/AFTER smart game;
  - not offer enough alternatives for selecting plausible wrong choices. For instance, if only one entity is detected in the story, the wrong choices (see discussion on "Phase C" above) are taken from *other* stories with different characters. However, since these choices are not very much plausible, the quality of the resulting smart game is reduced.

As a consequence, the manual revision phase takes a crucial importance in the whole generation process. Understanding the amount of manual effort is then of a major interest (Sec. 2), since it indicates the overall effectiveness of the automatic approach used in TERENCE and may give better insight in the tasks that has to be carried out as priorities (Sec. 3).

2 Manual Revision

The objective of the manual revision was to control the automatic generation and ensure the formal, technical, and content correctness of all the components of the game proposed for the stimulation of understanding of the written text. Details on the guidelines and on the software system that supported the manual revision can be found in [3].

The revision work was divided into 3 steps:

- Formal revision, i.e.: correction of grammatical and syntactic errors in the text, correction of punctuation, correction of the verb (present tense, active form), correction of referring expressions, check of sentence length and structure;
- Substantial technical revision, i.e.: check of game identification number, correction of the questions texts, correction of the solutions (by keeping fixed the main event), selection of new fixed events for solutions;
- Construction of cause/effect games, i.e.: text proposal, check out of proposals, games uploading.

Each operator studied the text of the story and reviewed all the games associated with it.

By proceeding on the basis of the output of the automatic generation, the manual review of TERENCE games was initiated through a software to speed up the audit work and its monitoring. The operator, accessing the software, could select the games by selecting the proper book and story. The Italian revision was conducted on 4 books: 3 books (16 stories) for students 7-9 years old and 1 book (9 stories) for students 9-11 years old.

The software shows on left of the interface the story text (so to let the operator to have the context always available) and enables in a tabular fashion to view the related games. After a game with its own ID was selected, it was possible to identify the type of game on the screen (before/after, before/while, before/while/after, what, while/after, who). Each game had its own *fixed event* (the event around which the entire game is build), completed with the event description, the game question, as well as a set of multiple-choice questions that depends on the selected game type.

The operator's job was firstly to verify the congruence between the fixed event text and the text of the proposed questions and solutions.

In general, the revision was structured, with the general principle to modify as little as possible the work of the automatic generation.

The revision consisted in correcting: (i) the grammatical and syntactic errors in the text; (ii) punctuation; (iii) verbs (present tense, active form); (iv) referring expressions; (v) sentence length and structure; (vi) the questions' texts [2]; (vii) the possible solutions; (viii) eventually the fixed event.

Each operator studied the story text, reviewed all the games associated with it and created from scratch the textual part of the causal games[3].

---

[2] With the aim of unambiguously identifying the event in the story. For instance, the same event text could be referred to different story episodes.

[3] The NLP module was in fact unable to detect any causal relation in the text.

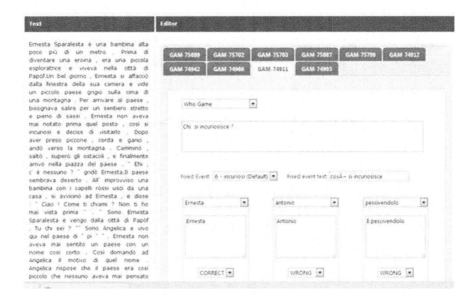

Fig. 3 Example of a review of WHO game (Book 2, Story 1)

In the revision of WHO-games, it sometimes happened that the solutions proposed were not consistent with the automatic generated question[4]. In such cases, it was necessary to change the solutions by choosing new entities from the drop-down menu and to guarantee the coincidence between entities and text. In the example shown in Fig. 3, the question asked is "Who is curious?". To make the necessary corrections, we had to (i) choose a new entity for each solution (by using the respective pull-down menu), and (ii) verify that it was the corret/wrong event actor.

We also had to take into account that each answer should be properly understood by all learners (poor comprehenders, deaf). Therefore, the changes were done by continually trying to work on subjects, preferring personal names, paying attention to the spatial distribution in the text and the kind of characters, and to not facilitate the reader in selecting the correct solution. Sometimes it was also necessary to completely change the game question associated with the fixed event, because the event was present several times in the text and associated with different subjects. In this case, it was also necessary to rewrite the question so to make it unambiguous.

In general, the review of temporal games was an even more challenging task, because it was necessary to locate the temporal coherence of proposed solutions on the basis of a fixed event, so that the wrong type solutions would exclude events with overlapping time intervals in to the right answers. For instance, the following corrections were applied to a before/after game of Book 2, Story 1 (Tab. 1).

---

[4] Probably due to errors introduced by the NLP anaphora resolution subsystem.

Table 1 Example pre/post revision for GAM 75702, Story 1, Book 2

Solution	Pre-revision	Post-revision
AFTER	to thank	The inhabitants of the land of "pf" thank Jasmine
BEFORE	Louis leads the electrician the wires	No change
WRONG	All manage to split the fairly rubbish without difficult calculations needs	All manage to divide garbage in the right way

Each operator had the task of filling in a diary (in a spreadsheet format, see Tab. 2), made up of 33 fields, all changes made in every revised game and depending on questions and solution types. This diary allowed the monitoring of all activities and their analysis as reported below.

Table 2 Revision games log

Items	Choices
Operator name	
Book number	
Story number	
Game ID	
Kind of game	Who, what, before/after, before/while, while/after, before/while/after
Did you have to change the main question of the game?	Yes/no
If yes, write the new main question text	
Did you have to change the fixed event?	Yes/no
If yes, write the new fixed event text	
If yes, write the old fixed event text	
Did you have to change the automatic solution?	Yes/no
Kind of solution	Wrong, correct, before, after, while, cause, effect
Did you have to change the event associated with the automatic solution?	Yes/no
If yes, write the new fixed event as in the menu	
If yes, write the new text as in the box solution	
Notes	

A total of 250 games were reviewed, with respect to 25 stories, with the highest proportion of games of type before/after (30%). Based on data reported by operators, the average review times were estimated. They were lower for reviewing a WHO-game, and higher for temporal games. On average, for a set of games, i.e. one game for each typology, it was necessary to work for circa 76 minutes. By considering the need for reading the related story, filling the excel, the average time of each operator to finalise a set of games was equal to circa 90 minutes, i.e. approx 15 minutes per game (see Tab. 3).

Table 3 Details about the revised games

GAME	n	%	Average time
WHO	25	10.00	10,6
WHAT	34	13.60	12
BEFORE/AFTER	74	29.60	12,8
BEFORE/WHILE	41	16.40	12,8
WHILE/AFTER	42	16.80	12,8
BEFORE/WHILE/AFTER	34	13.60	14,8
Total	250	100.00	75,8

Only in the 6% of all cases, it was necessary to change the automatically generated fixed event. In 72% of games the text of the event was corrected. The total number of changes (of both entities or choice events) was 120. The changes were necessary especially for the wrong choices, with 54 total changes (Fig. 4).

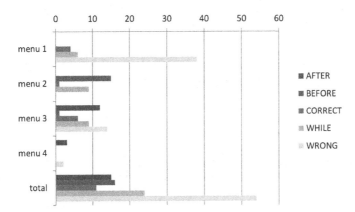

Fig. 4 Distribution of changes

The work of developing manual causal games was instead longer, as the operator had to invent the game by directly using, when possible, one of the fixed events already present in the story games. Overall, 75 causal games were created (both cause, effect, and cause/effect games). The average time spent for their development was equal to 23 minutes per game. The total work, also including game loading and final review, was about 30 minutes for each causal game.

3 Conclusion

Most of the revisions were of a formal nature, e.g. correct the letters for the names, verbs tenses. The main effort was in connection with reviewing the text: the phrases generated were incomplete or inconsistent, so it was necessary to continue to work on accents, the verb tenses and sentence length. A quantitative revision analysis showed a good level of automatic generation: in only 15 cases it was necessary to change the fixed automatically generated event. Mostly wrong type solutions were changed, with 54 change made necessary primarily to ensure a consistent level of difficulty in finding this solution.

Therefore, important and crucial improvements shall be directed towards the annotations of causal links (absent in this release), the text generation task, and the heuristic that takes care of the distractors (i.e., the wrong choices, and especially when the annotations are poor).

Acknowledgements. This work was supported by the TERENCE project, funded by the EC through FP7 for RTD, ICT-2009.4.2.

Special Thanks to the Team: A. Carbonelli, M.R. Cecilia, G. Cofini, G. D'Ascanio, M. Palmerio, J. Ranieri, M. Santucci, G. Sementilli, F. Zazzara.

References

1. Cofini, V., Di Giacomo, D., Di Mascio, T., Vittorini, P.: Evaluation Plan of TERENCE: when the User-centred Design Meets the Evidence-based Approach. In: Proc. of the ebTEL 2012 Workshop co-located with PAAMS 2012. Springer (2012)
2. Cofini, V., de la Prieta, F., Di Mascio, T., Gennari, R., Vittorini, P.: Design Smart Games with requirements, generate them with a Click, and revise them with a GUIs. Advances in Distributed Computing and Artificial Intelligence Journal 1(3), 59–68 (2012)
3. Cofini, V., Di Mascio, T., Gennari, R., Vittorini, P.: The TERENCE smart games revision guidelines and software tool. In: Vittorini, P., Gennari, R., Marenzi, I., Di Mascio, T., De la Prieta, F. (eds.) 2nd International Workshop on Evidenced-Based TEL. AISC, vol. 218, pp. 17–24. Springer, Heidelberg (2013)
4. De La Prieta Pintado, F., Di Mascio, T., Gennari, R., Marenzi, I., Vittorini, P.: Playing for Improving the Reading Comprehension Skills of Primary School Poor Comprehenders. In: 1st International Workshop on Pedagogically-driven Serious Games (September 1, 2012)
5. De La Prieta Pintado, F., Di Mascio, T., Gennari, R., Marenzi, I., Vittorini, P.: The TERENCE Smart Games: Automatic Generation and Supporting Architecture. In: 1st International Workshop on Pedagogically-driven Serious Games (September 1, 2012)
6. Gennari, R.: Deliverable 4.1: M.State of the Art and Design of Novel Intelligent Feedback Deliverable 4.1. Tech. rep., TERENCE project (2011)
7. Gennari, R.: Deliverable 4.2: Automated Reasoning Module. Tech. Rep. D4.2, TERENCE project (2012)
8. Hedberg, S.R.: Executive Insight: Smart Games: Beyond the Deep Blue Horizon. IEEE Expert 12(4), 15–18 (1997)
9. Moens, S.: Deliverable 3.1: State of the Art and Design of Novel Annotation Languages and Technologies. Tech. Rep. D3.1, TERENCE project (2012)
10. TERENCE Consortium: TERENCE web site, http://www.terenceproject.eu

Pedagogy-Driven Smart Games for Primary School Children

Fernando De la Prieta, Tania Di Mascio, Ivana Marenzi, and Pierpaolo Vittorini

Abstract. TERENCE is an FP7 ICT European project, highly multi-disciplinary, that is developing an adaptive learning system for supporting poor comprehenders and their educators. Their learning materials are stories and games, explicitly designed for classes of primary schools poor comprehenders, where classes were created via an extensive analysis of the context of use and user requirements. The games are specialised into smart games, which stimulate inference-making for story comprehension, and relaxing games, which stimulate visual perception and which train the interaction with devices (e.g., PC and tablet PC). In this paper we focus on how we used the pedagogical underpinnings and the acquired requirements to design the games of the system.

Keywords: Formalizations of pedagogical theories, serious games, game frameworks.

1 Introduction

More and more young children turn out to be poor (text) comprehenders: they demonstrate text comprehension difficulties related to inference-making skills, despite proficiency in word decoding and other low-level cognitive skills. Deep text

Fernando De la Prieta
University of Salamanca, Department of Computer Science, Spain
e-mail: fer@usal.es

Tania Di Mascio
University of L'Aquila, DISIM, Italy
e-mail: tania.dimascio@univaq.it

Ivana Marenzi
University of Hanover, L3S, Germany
e-mail: marenzi@l3s.de

Pierpaolo Vittorini
University of L'Aquila, MeSVA, Italy
e-mail: pierpaolo.vittorini@cc.univaq.it

P. Vittorini et al. (Eds.): *2nd International Workshop on Evidence-Based TEL*, AISC 218, pp. 33–41.
DOI: 10.1007/978-3-319-00554-6_5   © Springer International Publishing Switzerland 2013

comprehension skills develop from the age of 7-8 until the age of 11, when children develop as in- dependent readers. Nowadays, there are several pencil-and-paper reading strategies for improving text reading comprehension, and specifically addressed to poor comprehenders, which could be delivered by an adaptive learning system (ALS), that is, a suite of functionalities designed to deliver, track, report on and manage learning content for specific learners [8][9].

TERENCE is a EU project -http://www.terenceproject.eu-- that aims at delivering the first ALS for enhancing the reading comprehension of poor comprehenders, building upon effective pencil-and- paper reading strategies, and framing them into a playful and stimulating environment. Learners are primary school poor comprehenders, hearing and deaf, older than 7.

The goal of this paper is to explain how the playing material and tasks of TERENCE are designed and developed on top of an extensive analysis of the requirements of the TERENCE learners. First, the paper sets the groundwork by presenting the pedagogical theory and approach followed in TERENCE in Sec. 2. Then it outlines the types of data gathered for characterising the TERENCE learners and the analysed effective reading strategies and interventions for the TERENCE learners; in Sec. 3 and 4 is explained how the design and development of the TERENCE games, in particular, stems from such knowledge. For space limitations, we focus on the playing material, that is, games and playing tasks.

For information concerning the reading material and tasks, see [6]. Moreover, the models for the learning material and learners of the system are described in [4], how the user centered design (UCD) was used for them is in [2], whereas some of the adaptation rules are outlined in [5]. The game design for all the TERENCE games is in [3] and, finally, the architecture for games and their automatic generation is outlined in [10].

2   The Pedagogical Underpinnings

The theoretical framework underpinning of TERENCE is grounded on the constructivist pedagogical approach [15], which is a theory of learning that focuses on students being engaged in "doing", rather than passively engaged in "receiving" knowledge. In other words, constructivism states that learning takes place in contexts. This approach is committed to the general view that (1) learning is an active process of constructing rather than acquiring knowledge, and (2) instruction is a process of supporting that construction rather than communicating knowledge [11]. Nevertheless, knowledge does not simply arise from experience, but is build through experience over the current knowledge structures. The educator is required to orchestrate all the resources needed and must guide students in training them how to teach themselves [17].

Scaffolding is offered to the learner as an adequate environment where to find adequate learning material, compelling learning tasks, templates, and guide for the development of cognitive skills [21]. The focus is shifting from the educator directed instruction to a learner centered approach: the learner is at the center of the learning process. This yields that the learning material and tasks should be adequate for each learner profile, and that the learner should be guided through the material and tasks so as to achieve the learning goal.

The goal of this research is to enhance the reading comprehension of poor comprehenders. In order to do so, TERENCE system has being developed following the evidence-based design (EBD) and UCD [16], by involving a relevant number of real learners in the project and educators as is depicted in Fig. 1.

Fig. 1 The UCD and EBD design process of the TERENCE smart games

Data have been collected and analysed through user centred design methods, and then filtered through evidence-based sieves. The strategies of the educators can be framed in the three stages of the hermeneutic cycle explained in [20]. In particular, the explanatory stage can be broken down into the following reading interventions, done in class, mainly using question-answering and drawing:

1. the entire text is discussed with the learners, analysing the vocabulary unknown to the learners and paraphrasing the text;
2. the story is broken down into a sequence of episodes, if possible referring to the story grammar, that is, the story setting, the initiating episode, the culminating episode, the resolving episode, and the final episode;
3. finally, the time, the space and the characters of the story episodes are analysed together.

All the aforementioned interventions were considered for writing the requirements for the TERENCE game design. Constraints of the project triggered a prioritisation of the requirements which led in that visual aids were selected mainly for their expected efficacy for the pedagogy plan, according to the available empirical evidence: they should guide the child to better recall and correlate the information acquired reading the story via adequate visual representations. The effective interventions relevant for the TERENCE design have thus been hierarchically organised in levels according to their main pedagogical goal:

1. **time:** interventions for reasoning about temporal relations between events of the story, purely sequential (before-after) or not (all the others);

2. **causality:** interventions concerning causal-temporal relations between events of the story, namely, the cause of a given event (cause), the effect (effect), or the cause- effect relations between two events (cause-effect);
3. **characters:** interventions concerning characters, namely, who is the agent of a story event (who), what does a character in the story (what).

The context of use was thoughtfully analysed and specified for characterising the users of TERENCE, and hence stirring the design of the entire system. In this manner, the learning material and tasks were designed so as to be adequate to the real TERENCE learners. TERENCE learning material is made of stories and games for primary school children. Stories are organised in books, and games are distinguished into relaxing games, for relaxing the learners, and smart games, for assessing and stimulating reading comprehension. Each story is related to a set of c.a 2 relaxing games, and a set of c.a 15 smart games.

The EBD practice of the experts responsible for the pedagogical plan requires three main learning tasks in relation to the learning material of the system: (i) reading stories; (ii) playing with smart games for stimulating inference-making about stories; and (iii) playing with relaxing games for relaxing and motivating the learners. The pedagogical goal of relaxing games is to stimulate visuo-perceptual skills [14], instead smart games are designed to stimulate the recall and the correlation of the information acquired while reading a story.

Both smart and relaxing games are effective to provide a playful environment. When learning takes place in a playful environment, learning involves the learner actively and it increases his or her motivation and engagement. For this reason, an accurate stimulation plan has been carefully designed according to various experts´ feedback. The experts, in view of their evidence-based experience with stimulation plans for children, suggested to focus on specific types of stimuli, namely, inference-making about events and their relations, in order to train the learners on this within the stimulation plan. Given that inference-making is the specific focus of the project, interventions were related to inference making about stories, and to deep text comprehension more in general.

The results of the stimulation plan setted specific requirements on the design of the TERENCE smart games, related to Time, Accuracy and Level of difficulty. As a result of these requirements, relaxing games and smart games have been designed as follows:

- *Relaxing games* are modeled on familiar games and serve to make the stimulation plan more appealing to the learner. Each type of relaxing game aims at stimulating a type of visio-perceptual interaction used in smart games that the TERENCE learners are unlikely to have [14]. Therefore, they relax and motivate the learners to use the TERENCE system within the stimulation plan, after or before playing with smart games.
- *Smart games* serve to guide the child to better recall and correlate the information acquired reading the story. Moreover, according to the performances of the learners over the smart games, the adaptive module can decide whether to move the learner from one story level to another. Therefore smart games lay at the core of the stimulation plan.

To summarise, according to the experts of the stimulation plan, TERENCE should not propose games concerning causes to children with very low reading comprehension skills, and should avoid games that are likely to be too unchallenging, like who-questions and what-questions, with children with good reading comprehension skills. In the case of children with very low skills, too demanding games can easily lead to frustration whereas, in the case of children with good skills, unchallenging games can easily lead to boredom.

3 Characterisation of Learners for Playing Tasks

By using the UCD, we extensively and deeply analysed the context of use and the learners´ requirements, thereby specifying classes of learners for the system. The learning material and tasks of the system were designed for those classes of learners. The first part of this section outlines the type of data collected and analysed for specifying the classes of users. The second part outlines the type of data collected and analysed for designing the learning material and the learning tasks.

3.1 Classes of Learners

The specific goal of the project distinguishes the two classes (deaf and hearing learners). These classes were refined on the basis of the results of the analysis of data for the context of use and user requirements. Such data have been gathered via a mix of expert-based method inquiries and user-based method inquiries. The learners involved were about 300 in Italy and about 300 in the UK; the educators involved were about 50 in Italy and about 30 in the UK. Learners are currently represented by five classes in Italy and four classes in UK [14]. The most significant features related to the characteristics of the users and considered for deriving the TERENCE classes are biographical information, personality and usage of technology. All the classes and the features used for deriving the TERENCE classes were then specified using personas, which are explained in [2] [14].

3.2 Playing Tasks

All data for the game requirements have been gathered through UCD methods, the results of which are reported in [18] as tasks. In particular, the data for relaxing games are popular causal video games which is a video-game meant for casual gamers who come across the game and can get into the gameplay almost immediately. This means that the causal game has usually simple rules and usually it can be played everywhere, anytime and with any device. The data for smart games are mainly diverse reading strategies by pedagogy experts working as therapists with poor comprehenders, cognitive psychologists or educators. The TERENCE smart games were then layered into similar levels as the previous interventions, that is, smart games at the entry level for reasoning about characters, games at the intermediate level for time, and games at the top level concerning causality.

The following section delves into how the design and development of the smart and relaxing games is carried out via the TERENCE framework.

4 The Design and Development of Games via TERENCE Framework

According to the game design guidelines [1] for specifying the gameplay of the TERENCE games we analysed the data for the gameplay of each TERENCE game, then we abstracted the common characteristics in the TERENCE game framework presented in Table 1.

Table 1 The TERENCE game framework

Name	name of the game	
Instructions	instructions concerning the game, for the learner: specific to the game instance; motivational; concerning the rules	
Choices	the choices available to the learner; their availability is state depen- dent	
Solutions	Correct	wrong
	which choices are correct solutions	which choices are wrong solutions
Consistency f	correct	wrong
	a yes-message for correct solutions	a no-message for wrong solutions
Explanatory f.	correct	wrong
	explanatory message for correct solutions	explanatory message for wrong solutions
Solution f.	a message consisting in the correct solution	
Smart points (e.g., coins)	$K.P(\theta)$, where θ is the underlying ability of the learner for the game, and K is a constant ranging over natural numbers	
Relaxing points (e.g., starts)	M, that is, a natural number from 1 up to N	
Avatar	the states of the avatar	
Time	resolution time t_r	
Rules	the rules for the game mechanics, specifying the states of the system, the learners' actions and the transitions from state to state through the learner's actions	

The framework serves to specify in a structured manner the above data for the gameplay of the TERENCE smart and relaxing games, essentially, through a timed transition system, with states of the system, and transitions labelled by the player's actions and time constraints.

In the following, we firstly present the common elements of framework for relaxing of smart games:

- The *instructions* for the game are questions specific to the game instance; of motivational type and usually related to the learner avatar; concerning the rules.

- The available *choices* may change from state to state of the game: at the beginning all the choices are available; when the play starts, some choices may become unavailable. The solutions for the game list the choices or their combination that form a correct solution to the game (correct), and those that do not (wrong).
- The *feedback* for the game is specialised into a consistency feedback (yes, no), an explanatory feedback for finding a correct solution (for correct) or for spotting what is wrong in the current solution (for wrong), and a solution feedback (the correct solution).
- The states of the *avatar* in the gameplay are of two kinds: happiness for the correct solution, disappointment for the wrong solution. The resolution time is a constant.

The *Smart points* are the points a learner with a specific reading comprehension level can gain in a smart game. These points can be calculated using the IRT [7], so that the more difficult a game is (assessed to be) for a learner, the more points the learner can gain in resolving correctly that game. Relaxing games have relaxing points instead of smart points. Relaxing points should be easy to cumulate, so as to motivate the learner to keep on playing and, in so doing, earning attributes for the avatar.

Now, like points, *rules* are different for smart games and relaxing games.

- **Smart rules**, the pedagogical plan establishes requirements for the actions that the learner can take, the states the system can be in, and constraints on them. In the following, we sketch the actions, the states and the constraints for smart games:

  The plan also recommends diverse types of feedback if the learner makes a wrong choice: first, a no-consistency feedback for signaling that the solution is wrong, and then an explanatory feedback are given.

  The plan also suggests a solution feedback if the leaner chooses no solution within the resolution time or the number of wrong solutions overcomes the wrong attempts limit.

  The main states the system can be in are as follows:

  (i)   the initial state, in which the learner score s and resolution time t are set to 0, the smart points for the learner are computed as a function of the learner ability in the game, all the choices are set as available, and the number of wrong answers is set to 0;

  (ii) a terminal state reachable via a correct action, in which a yes-consistency feedback is given, the score is displayed and the avatar is in the happy status;

  (iii) a terminal state reachable via a skip action, in which the solution feedback is given, the null score is displayed and the avatar is in the displeased status;

  (iv) a state, reachable via a wrong action, in which a no-consistency feedback is given, an explanatory feedback is given, the set of available choices is updated, and the number of wrong answers is updated; and

  (v)   a terminal state reachable via a wrong action, in which the no-consistency feedback is given, the solution feedback is given, the null score is displayed and the avatar is in the displeased status.

- **Relaxing rules** have the same rules as well, based on common rules for casual games.

    In the initial state, the score and resolution time are set to 0. From any non-terminal state, we can have the following: let N be the number of relaxing points that can be cumulated in a relaxing game, if the score is less than N and, within the game's resolution time, the learner chooses a correct solution, then the system shows the yes-consistency feedback, and the score gets increased by 1. But, if the learner chooses a wrong solution, then the system shows the no-consistency feedback, the game terminates and the system shows the disappointed avatar; otherwise, the system terminates the game, shows the score and the happy avatar.

4.1 Conclusion and Game Prototype

In this paper we explained how the playing material and tasks of TERENCE are designed and developed on top of an extensive analysis of the requirements of the TERENCE learners based on (EBD) and UCD. This study finalizes with the development of the prototypes of smart games, like the one in Fig. 2. The development procedure, from the framework via the visual template to the prototypes is reported in [3].

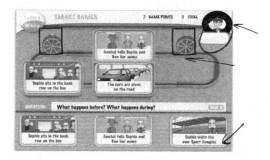

Fig. 2 An instance of a prototype of a before-while smart game

Acknowledgements. The authors' work was supported by TERENCE project, funded by the EC through the FP7 for RTD, Strategic Objective ICT-2009.4.2, ICT, TEL. The contents of the paper reflects only the authors' view and the EC is not liable for it. Gennari work was also funded through the CRESCO and DARE projects, financed by LUB and the Province of Bozen-Bolzano.

References

1. Adams, E.: Fundamentals of Game Design. New Riders (2010)
2. Alrifai, M., de la Prieta, F., Di Mascio, T., Gennari, R., Melonio, A., Vittorini, P.: The Learners' User Classes in the TERENCE Adaptive Learning System. In: Proc. of the ICALT 2012 (2012)

3. Alrifai, M., Gennari, R.: Deliverable 2.3: Game Design. Tech. Rep. D2.3, TERENCE project (2012)
4. Alrifai, M., Gennari, R., Tifrea, O., Vittorini, P.: The user and domain models of the TERENCE adaptive learning system. In: Vittorini, P., Gennari, R., Marenzi, I., de la Prieta, F., Rodríguez, J.M.C. (eds.) International Workshop on Evidence-Based TEL. AISC, vol. 152, pp. 83–90. Springer, Heidelberg (2012)
5. Alrifai, M., Gennari, R., Vittorini, P.: Adapting with evidence: The adaptive model and the stimulation plan of TERENCE. In: Vittorini, P., Gennari, R., Marenzi, I., de la Prieta, F., Rodríguez, J.M.C. (eds.) International Workshop on Evidence-Based TEL. AISC, vol. 152, pp. 75–82. Springer, Heidelberg (2012)
6. Arfe, B.: Deliverable 2.2: Repository of Stories. Tech. Rep. D2.2, TERENCE project (2012)
7. Baker, F., Seock-Ho, K.: Item Response Theory. Dekker Media (2012)
8. Brusilovsky, P.: Adaptive hypermedia: From intelligent tutoring systems to web-based education. In: Gauthier, G., VanLehn, K., Frasson, C. (eds.) ITS 2000. LNCS, vol. 1839, pp. 1–7. Springer, Heidelberg (2000)
9. Brusilovsky, P., Karagiannidis, C., Sampson, D.: Layered evaluation of adaptive learning systems. International Journal of Continuing Engineering Education and Life-long Learning 1(14), 402–421 (2004)
10. De la Prieta, F., Di Mascio, T., Gennari, R., Marenzi, I., Vittorini, P.: The TERENCE Smart Games: Automatic Generation and Supporting Architecture. In: Proc. of the PDSG 2012 Workshop co-located with EC-TEL 2012. CEUR-WS (2012)
11. Duffy, T.M., Cunningham, D.J.: Constructivism: Implications for the design and delivery of instruction. In: Jonassen, D.H. (ed.). Simon & Schuster Macmillan, New York (1996)
12. Mascio, T.D.: First User Classification, User Identification, User Needs, and Usability goals, Deliverable D1.2. Tech. rep., TERENCE project (2012)
13. Nanjappa, A., Grant, M.M.: Constructing on constructivism: The role of technology. Electronic Journal for the Integration of Technology in Education 2 (2003)
14. Norman, D.: The design of everyday things. Doubleday, New York (1998)
15. Rizzo, R.: Multimodal and Multimedia aspects of English Language Teaching and Studies in Italian Universities. Ibis, Como (2009)
16. Slegers, K., Gennari, R.: Deliverable 1.1: State of the Art of Methods for the User Analysis and Description of Context of Use. Tech.Rep.D1.1, TERENCE project (2011)
17. Valeriani, A.: Ermeneutica retorica ed estetica nell'insegnamento verso l'oriente del testo. Andromeda (1986)
18. Yelland, N., Masters, J.: Rethinking scaffolding in the information age. Computers and Education 48, 362–382 (2007)

Technology and Elaboration of Information in Aging: Preliminary Data of Verbal vs. Visual Performance

Dina Di Giacomo, Massimiliano Palmiero, and Domenico Passafiume

Abstract. Recently, the researchers' interests were focused on the interaction between old age and e-learning, with the aim of investigating the critical factors that might determine the successful aging. In order to verify the modifications of information elaboration competence in aging though Information Technology. We submitted 125 subjects, divided in 5 groups on the basis of subjects' age, to experimental verbal and visuoperceptual tests, both computerized. The results showed that on visuoperceptual test the Young group was faster than Junior group, whereas On verbal test the young group was slower than adult group. These results highlighted that the successful in the use of e-learning may be influenced by cognitive and emotional factors in aging.

1 Introduction

E-learning is the use of Internet technologies to enhance knowledge and performance. It is emerging as the new paradigm of modern education. Successful e-learning takes place within a complex system involving the student experience of learning, teachers' strategies, teachers' planning and thinking, and the teaching/learning context. Several researches was conducted on a) the e-learning impact in developmental age, b) the influence of technologies on the educational programs, c) the use of information technology at the work Creswell et al. [1] and Wang et al. [4]. Moreover, the use of e-learning and its successful may be influenced by critical factors as the learner satisfaction. Sun et al. [5] were examined the critical factors affecting learners' satisfaction in e-learning. Their results evidences some critical factors: learner computer anxiety, instructor's attitude toward e-Learning, e-Learning course flexibility, e-Learning course quality, perceived usefulness, perceived ease of use, and diversity of assessments. Many researchers

Dina Di Giacomo · Massimiliano Palmiero · Domenico Passafiume
Department of Life, Health & Environmental Sciences, University of L'Aquila, Italy
e-mail: {dina.digiacomo,domenico.passafiume}@cc.univaq.it,
        massimiliano.palmeiro@univaq.it

P. Vittorini et al. (Eds.): *2nd International Workshop on Evidence-Based TEL*, AISC 218, pp. 43–48.
DOI: 10.1007/978-3-319-00554-6_6   © Springer International Publishing Switzerland 2013

have focused the e-learning impact on the quality of life on childhood and on adulthood population. Few studies were conducted on the e-learning and its usability in ageing. Githens et al. [3] argued on the opportunities and barriers for the olders adults in the e-learning: they described the changing notion of work and learning in older adulthood, the prejudices about older adults' use of technology, the types of e-learning programs for older adults (i.e., programs for personal growth and social change, workforce development, and workplace learning), and the barriers to older adults' full participation in e-learning. These evidences highlighted the necessity to improve the knowledge on the different approaches to the e-learning in life span. The question isn't only the implementation of information technologies applications and features but the interaction among the information program and aging target to consent an adaptation behavioral to increase the successful use of e-learning.

Aim of the study is to evaluate the relation between cognitive abilities in aging and use of information technologies. We want to examine the performance of subjects in different range of age on verbal and visual performance and to analyze the variables that influence the different level of performance; moreover, we might highlight weaker age in the learning and contemporary we might evidence the strenght age in the use of technology.

2 Materials and Methods

2.1 Subjects

The sample involved in the study was composed of 125 native Italian speaker subjects, balanced for sex and education, age ranging 20 to 70 years olds, divided in 5 groups: 1) Young Group (YG) composed of 25 subject age ranging 20-29 years , mean age 23,9 (sd ±5,1); 2) Junior Group (JG) composed of 25 subject age ranging 30-39 years, mean age 35,8 (sd ±3,2); 3) Adult Group (AG) composed of 25 subject age ranging 40-49 years , mean age 46,5 (sd ±2,5); 4) Senior Group (SG) age ranging 50-59 years , mean age 54,2 (sd ±4,8); Elderly Group (EG) age ranging 60-69 years, mean age 66,0 (sd ±3,0). All participants reported negative story of psychiatric or neurological disease or alcohol and drug problems. All subjects signed written informed consensus.

2.2 Testing

The psychological battery was composed of 2 tasks aimed to evaluate verbal and visual performance of the sample. The Tasks were developed from the Semantic Associative Test [2] as the verbal and visual version. The experimental test was carried out in 'paper and pencil' form. In this experiment we modified the test: we prepared an PC version. The test required to touch on the screen the word/figure better associated to the figure/word-target between 3 choices. Each test have the same items. In Figure 1 was reported example item tests.

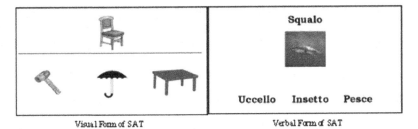

Visual Form of SAT   Verbal Form of SAT

Fig. 1 Example of items of the SAT: visual and verbal versions

In the experiment we applied the tests in computerized version. We want examine the performance of the sample analyzing the accuracy and the execution time, so we elaborate the tests from 'paper and pencil' version (standard) to computerized version.

Experimental condition. In Figure 2 is represented the experimental condition. The subject was in front of the PC touch screen (screen 32"; resolution 1920 x 1080; distance subject-monitor = 70 cm); the subject's right hand was placed on the table close to an external touchpad (208 x 138 x 7,5 mm; active area 125 x 85 mm); When the item appeared on the screen, the subject had to touch fast first the pad and then the PC screen. The warning sound of 500 Msec precedes the appearance stimulus. The stimulus remained on the screen until the subject answers touching the screen. The task execution modality was in 3 steps: 1 = relaxing state; 2 = touch of the pad for decision time performance; 3 = touch of the screen for accuracy performance. In the relaxing state the item appeared on the screen and the timer was trigged. As soon as the subject had decided the correct choice, he had to touch the pad: the time in between the appearance of the item and the touch of the pad was considered the decision time; finally, the touch on the PC screen of the correct/wrong choice was considered the response accuracy. The participants were verbally instructed on the task, and were submitted to three practice items.

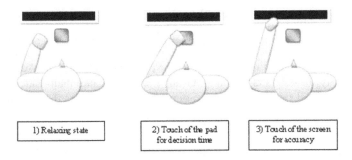

| 1) Relaxing state | 2) Touch of the pad for decision time | 3) Touch of the screen for accuracy |

Fig. 2 The execution condition in experimental test

2.3 Procedure

The participants were adult voluntaries recruited from the University population and City Clubs. Each participant was tested in a quiet room reserved for testing.

The test administration time was approximately 40 minutes. The examiners were Psychologists, graduate students from the School of Clinical Psychology of the University of L'Aquila, Italy. The examiners were been blind to the research objectives.

3 Results

The data have been analyzed by the computerized program 'Statistica'. The significance level fixed was alpha <0.05.

Data submitted to statistical analysis were both accuracy and decision time.

First of all, we conducted a statistical analysis on the accuracy data. A repeated measures ANOVA was showed no significative difference between the groups and tests. This data was expected because we want' verified the efficiency in the semantic ability.

A statistical analysis was conducted on the decision time (DT) in msec. The Table 1 reports the means and standard deviation.

Table 1 Means of time decision and deviation standard of sample in verbal and visual performance

Sample	TIME DECISION Verbal performance	TIME DECISION Visual performance
YOUNG GROUP	9,1 (±2,7)	7,4 (±2,4)
JOUNIOR GROUP	8,9 (±2,7)	8,6 (±3,1)
ADULT GROUP	6,2 (±2,1)	8,1 (±2,7)
SENIOR GROUP	9,4 (±3,0)	8,8 (±2,8)
ELDERLY GROUP	11,3 (6,9)	8,0 (±3,7)

An 2x5 ANOVA showed significative difference between tasks neither between groups; no significative resulted the interaction groups x performance. The Post-Hoc Analysis (LSD test) evidenced in verbal performance the Adult Groups significatively faster than Young Group (<0,02), and than Elderly Group (<0,008); in visual performance, the Young Group was faster than Junior Group (<0,02).

4 Discussion and Conclusions

Our comparative study was focused on the performance (accuracy and response time) of subjects with different age in order to study the influence of age on the cognitive elaboration of information using information technology. Particularly, our study wants to verify the ability to use the touch technology in verbal and visual modality during aging, analyzing the performance of the sample (20-70 years range of age). The results showed different ability to elaborate the information: in the visuoperceptual test, the Young Group was faster than Junior Group; in the Verbal Test the Young Group was slower than Adult one. The lack of statistical significant difference between groups and tasks as for accuracy, rule out the role of the semantic abilities in the time differences. The presence of differences between groups in the time measure seem to indicate that the ability of use the information technology depend not only on the age but on modality of task presentation and of execution, too

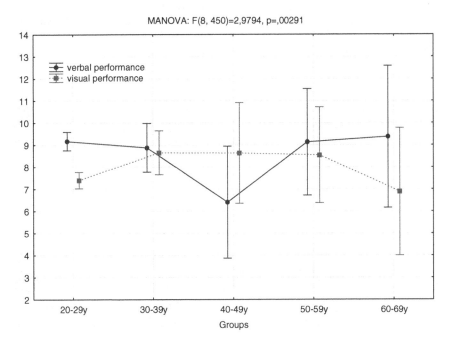

Fig. 3 Representation of the verbal and visual performance at the SAT of the 5 groups. The Y values represent the Msec time.

According to Githens et al. [3], the e-learning in older adults results difference that of younger and juniors and these difference, probably depend by the interrelation among cognitive and emotive conditions. We suggest that the emotional critical factors to determine the successful e-learning performance in adulthood could be related to the self-consciousness of different modality and speed of elaboration information (verbal vs. visuospatial abilities). It's also to keep in mind, the information technology and its applications were early programmed preeminent for younger population. Recently, the research interesting was focused also on the influence of the e-learning on quality life in aging and its improvement.

References

1. Creswell, J.W.: Educational research: planning, conducting and evaluating quantitative and qualitative research. Prentice Hall, Upper Saddle River (2005)
2. Di Giacomo, D., De Federicis, L.S., Pistelli, M., Fiorenzi, D., Sodani, E., Carbone, G., Passafiume, D.: Loss of Conceptual Associations in Mild Alzheimer's dementia. Journal of Clinical and Experimental Neuropsychology 34, 643–653 (2012)
3. Githens, R.: Older adults and e-learing: opportunities and barriers. Quarterly Review of Distance Education 8, 329–339 (2007)
4. Wang, M., Ran, W., Liao, J., Yang, S.: A performance-oriented approach to e-learning in the workplace. Educational Techonology & Sociaty 13, 167–179 (2010)
5. Chen, S.-P., Tsai, R.J., Finger, G., Chen, Y.Y., Yeh, D.: What drives a successful e-Learning? An empirical investigation of the critical factors influencing learner satisfaction. Computers & Educations 4, 1183–1202 (2008)

Use of Flight Simulators in a Flight Mechanics Course: Potentials and Weaknesses

Diego Domínguez, Deibi López, and Jesús Gonzalo

Abstract. This paper is focused on the process of acquiring and developing knowledge in flight mechanics by means of the use of flight simulators. The analysis of student improvements shows reinforcement in their previous knowledge and quick acquisition of new concepts; however, there is place to much more enhancements taking advantage of these high stimulating exercises and trying to avoid student distractions during the process.

Keywords: flight simulation, interactive learning, flight mechanics.

1 Introduction

New teaching methods, and especially the ones related to learning technologies, are demanded in European Higher Education Area (EHEA). In recent years, it has been pointed that modelling environments can supply effective pedagogical strategies dealing with complex real-world systems [1]. Edgar Dale's Cone of Experience [2] points out that direct experience represents the greater depth of learning; from this perspective, the participation of students in virtual environments, where they have to implement their learning, implies a reinforcement of the teaching-learning experience. Flight simulators, good examples of what is known as "serious games", can promote the understanding, integration and application of concepts and improve performance, as other alternative to traditional lectures.

Although flight simulation sessions are at the top level of airline pilot training, it is almost an emerging field at university in engineering courses, with few written references [3, 4]. In this context, the use of flight simulators should be considered an interesting way of acquiring competences on flight mechanics as it involves the understanding of important facts about physics of flight and aircraft performance. Moreover, a successful flight also requires a kind of sensitivity in order to understand and anticipate the impact of the control inputs on aircraft situations [4]. Because of these reasons, the use of flight simulators, safe and moderately affordable, has been considered in this work to enhance the learning process on the subject.

Diego Domínguez · Deibi López · Jesús Gonzalo
Aerospace Engineering Area, Universidad de León, León, Spain
e-mail: {ddomf,dlopr,jesus.gonzalo}@unileon.es

P. Vittorini et al. (Eds.): *2nd International Workshop on Evidence-Based TEL*, AISC 218, pp. 49–55.
DOI: 10.1007/978-3-319-00554-6_7   © Springer International Publishing Switzerland 2013

The study of the effectiveness of this technology-based learning, and how such novel educational techniques are influencing students during the learning process, needs to be measured in order to ensure the ability of the method, being this the main purpose of the present work.

2 Methodology

The experiment was conducted at the Universidad de León during December 2012. All the students who take part on it were enrolled in the Aerospace Engineering degree program and coursing Flight Mechanics (3rd year students). The whole experiment comprised two different parts, the first one was a single test including questions about flight mechanic issues. The second part was the exercise with the flight simulator. The test was executed twice, initially prior to the practice and again when the practice was concluded. All students in the class (41) were invited to participate although the final number of participants in the practical sessions was 31, distributed in two groups, allowing us to create a 10 people control group. All the students, including those in control group, made both tests.

2.1 The Equipment

The practical lessons were developed at the flight simulators belonging to the *Centro de Simulación Aérea de la Universidad de León* (Saule). The simulator has a virtually real aircraft cockpit reproducing a Beechcraft Baron 58 with X-Plane as a simulation engine. X-Plane is a realistic flight simulator capable of simulate aircraft performance and weather effects. It has good models of airplanes, which in some versions are already approved by Federal Aviation Administration (FAA) for full motion simulator to train commercial airline pilots.

Fig. 1 Left: external view of the flight simulator. Right: cockpit view

2.2 The Practical Session

We have developed a complete set of educational simulations, where the students and the learning process are the centre. Those simulations are at the same time instructive, motivating and fun, allowing students to explore an important range of flight situations and maneuvers. These practical lessons are described next.

Exercises has been designed in order to test several commands and responses of the plane, which are: climb and descent, stabilize different slip flights, symmetrical turns, airplane stalls with three different flap-configurations (0%, 50% and 100%), phugoid mode, spiral mode and gliding flight. Every pair of students has 10 minutes to do the simulation distributed as follow: first of all, each student has 3 minutes to become familiar with the aircraft controls and responses, with the objective of stabilize the plane and achieve cruise flight. Then, each one has to do one of the listed exercises. After this 10 minutes the lecturer give some guidelines and key concepts about the completed practices, asking some interactive questions to the students, both the pair who has made the flight and the rest of students.

3 Results and Discussion

Test results before and after the practical sessions are showed and commented in this section. The questions are distributed in three groups, according with the achieved results, and followed by a short discussion. Each question includes a graph with statistics from the test, left column are students who made the practice (results from students belonging to both practice groups are showed together) and right column are the answers of the control group that did not carry out the simulations. On the other hand, first row are results from the test made before the simulations and the second one corresponds to test completed after.

Table 1 Questions with high improvement rates

N° 1 **Which one is correct ¿if an engine fails in a twin-engined plane during the cruise and no corrective action is applied?** a) It produces slip b) It produces roll c) It produces pitch **Correct answer:** a)	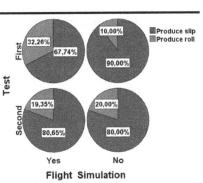
N° 3 **During an horizontal equilibrium flight the flaps are somehow deployed and the pitch angle remains zero, the aircraft:** a) Increase flight altitude b) Reduce flight altitude c) Fluctuates up and down **Correct answer:** a)	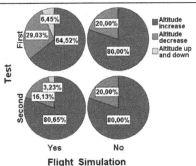

Table 1 (*continued*)

Nº 4

If only roll is permitted, are coordinated turns possible?
a)   No, the use of the pedals is needed
b)   No, the use of the elevator is needed
c)   No, both are needed
d)   Yes.

Correct answer: c)

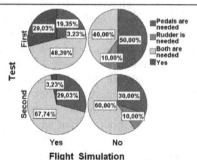

Nº 6

If the ball gauge is right-displaced from is neutral position doing a horizontal right turn:
a)   The pilot must push the right pedal to correct
b)   The pilot must push the left pedal to correct

Correct answer: a)

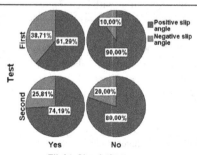

Nº 7

If the ball gauge is right-displaced from is neutral position doing a horizontal right turn:
a)   The slip angle is positive in body axes
b)   The slip angle is negative in body axes

Correct answer: a)

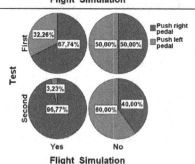

Nº 8

In order to carry out a turn with constant angular speed, the elevator angle is:
a)   Larger than the necessary for horizontal, uniform and rectilinear flight (HURF) at same speed
b)   Lower than the necessary for HURF at same speed
c)   Same as necessary for HURF at same speed

Correct answer: a)

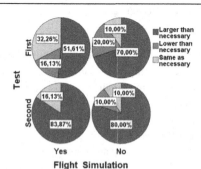

1 - The practice has increased the understanding of the students about aircraft performance, as it included flight situations with slip flight and engine failure.

3 - It is clear that the practice has improved students' knowledge about aircraft flight envelope with deployed flaps. The effect of this action could be easily appreciated during the stall practice, when the flaps are deployed by the students.

4 - The question is related with causes and consequences of coordinated turns, which are recorded in the topics of the course. The obtained results are favorable.

6 - This is a typical example of improvement. The conducted practices allow the student to get better knowledge about the causes and consequences between the commands and the plane response.

7 - The student can check the instruments in real time with the commands, which allows him to support his theoretical knowledge. This is a good example of improvement due to flight simulation practices.

8 - The rate of improvement of students having flown the simulator is much higher that the observed in the control group, as the answer can be easily derived from the practical experience and the piloting feeling.

Table 2 Questions with reduced improvement rates

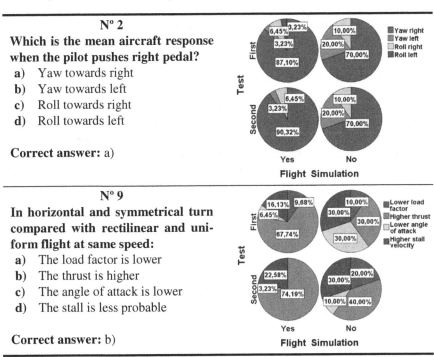

| N° 2 |
| Which is the mean aircraft response when the pilot pushes right pedal? |
| a) Yaw towards right |
| b) Yaw towards left |
| c) Roll towards right |
| d) Roll towards left |
| Correct answer: a) |

| N° 9 |
| In horizontal and symmetrical turn compared with rectilinear and uniform flight at same speed: |
| a) The load factor is lower |
| b) The thrust is higher |
| c) The angle of attack is lower |
| d) The stall is less probable |
| Correct answer: b) |

2 - Most of the students knew the correct answer before the practice and only a small proportion of them improved their marks. The result is surprising as some of the practical exercises specifically required pushing the pedals.

9 - The erratic behavior of control group opposes to the good response of students after the simulated flights. The correct solution is more frequent and, besides, nonsense responses have disappeared, being this a relevant result.

Table 3 Questions without improvement

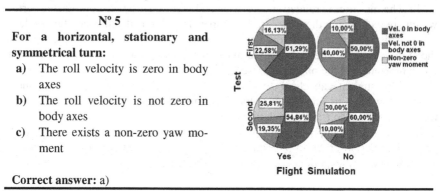

N° 5
For a horizontal, stationary and symmetrical turn:
a) The roll velocity is zero in body axes
b) The roll velocity is not zero in body axes
c) There exists a non-zero yaw moment
Correct answer: a)

5 - In this case the expected results were not fully reached as the increase in good answers is low. The time of the practice may be not enough to assimilate some positioning and stability concepts which cannot be easily appreciated with simulators.

4 Conclusions

A correct interpretation of student feedback –in this case by means of the test– is critical in order to ensure the success of the educational activity [5]. Students were really receptive in order to take part in the activity, proving flight simulation attractiveness. However, it is important to look for objective indicators that could demonstrate an improvement in the student knowledge because, other way, it only could be considered an entertainment activity. Some important results are summarized below.

Most flight variables were recorded for each flight; their visualization shows that the accomplished maneuvers are not strictly well executed from a piloting point of view, as they do not produce 'clean trajectories'. However, it is important to point out that this was not relevant for the learning process (Fig. 2). This conclusion is really important, as it ensures that any student can learn by means of flight simulation without being a trained pilot, making this methodology suitable for a university course whose main purpose is not train pilots but engineers.

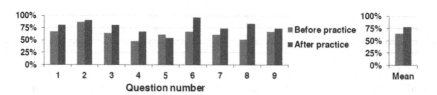

Fig. 2 Percentage of correct answers for each question before and after the practical lessons

Higher marks after the flight ensure that student knowledge has been improved. Although there are differences from one question to other, the average improvement percentage for the overall test was 14%. We believe there is room for better performance attending to the *distraction hypothesis* [6]: entertainment distracts for learning, especially those students who had not had previous contact with flight simulators. In other words, the students are not thoughtful enough to take full advantage of the potential knowledge improvement that the experience offers and it is difficult to maintain an adequate attention level.

In order to minimize this kind of distractions two new ideas are proposed for the next course: to extend the time students spend in the simulator, allowing some playing time before tests, and to revisit the exercises looking for a better match with the concepts to learn.

Acknowledgements. We would like to express our gratitude to the *Instituto de Automática y Fabricación* (IAF) of Universidad de León for granting the use of the flight simulators and to Everton Spuldaro for his relevant technical assistance.

References

1. Tarantino, G., Fazio, C., Sperandeo-Mineo, R.M.: A pedagogical flight simulator for longitudinal airplane flight. Comput Appl. Eng. Educ. 18, 144–156 (2010)
2. Dale, E.: Audio-visual methods in teaching, 3rd edn. Rinehart & Winston, New York (1969)
3. Brodeur, D.R., Young, P.W., Blair, K.B.: Problem-based learning in Aerospace Engineering Education. In: Proceedings of the 2002 American Socierty for Engineering Education (2002)
4. Valdés, R.A., Sanz, L.P., Alonso, J.F.: The use of flight simulators for airspace design in engineering education. Int. J. Eng. Educ. 4, 757–767 (2011)
5. Mayer, I.: Towards a Comprehensive Methodology for the Research and Evaluation of Serious Games. Procedia Comput. Sci. 15, 233–247 (2012)
6. Ritterfeld, U., Weber, R.: Video Games for Entertainment and Education. In: Vorderer, P., Bryant, J. (eds.) Playing Video Games. Motives, Responses and Consequences. Lawrence Erlbaum Associates, Mahwah (2006)

Supporting Context-Aware Collaborative Learning Activities by CAFCLA

Óscar García, Ricardo S. Alonso, Dante I. Tapia, and Juan M. Corchado

Abstract. The integration of Information and Communication Technologies (ICT) in daily life has improved the learning process by means of context-aware technologies. Through the use of technology, new ways of learning has emerged allowing to become the learning process more ubiquitous. However, it is necessary to develop new tools that can be adapted to a wide range of technologies and different application scenarios. This paper presents CAFCLA, a framework that allows developing context-aware learning applications. CAFCLA integrates different context-aware technologies, so that learning applications designed, developed and deployed upon it are dynamic, adaptive and easy to use by users such as students and teachers.

Keywords: Computer Supported Collaborative Learning, Context-Aware Learning, Real Time Location Systems, Wireless communication.

1 Introduction

The society has been flooded with a wide range of different mobile devices which user interfaces or communication skills are improved day by day [1]. Thanks to these advances, we are surrounded by technology that has changed our habits and customs [2] by receiving information from multiple sources. The ability of the devices to collect, offer and react to the information from the environment is the main topic in Context-aware Computing [3].

The usage of Information and Communication Technologies (ICT) has been present in educational innovations over recent years [4], modernizing the traditional transmission of contents through electronic presentations, email or more complex learning platforms such as Moodle and fostering collaboration between

Óscar García · Ricardo S. Alonso · Dante I. Tapia · Juan M. Corchado
Department of Computer Science and Automation, University of Salamanca.
Plaza de la Merced, s/n, 37008, Salamanca, Spain
e-mail: {oscgar,ralorin,dantetapia,corchado}@usal.es

P. Vittorini et al. (Eds.): *2nd International Workshop on Evidence-Based TEL*, AISC 218, pp. 57–65.
DOI: 10.1007/978-3-319-00554-6_8   © Springer International Publishing Switzerland 2013

students (Collaborative Learning) [5]. Beside the use of those general-purpose tools in education, other tools that make more specific use of technology have appeared. This applies to those that make use of Context-awareness information and ubiquitous computing and communication.

The inclusion of context-awareness in educational scenarios and processes refers to Context-aware Learning [6], a particular area of application of Context-aware Computing [3]. Moreover, the ability to characterize and customize the context that surrounds a learning situation at a certain time and place provides flexibility in the educational process. This way, learning does not only occur in classrooms, but also in a museum, park or any other place [7], obtaining ubiquitous learning spaces. Thus, there is an extensive literature that addresses the problem of this kind of learning, highlighting those works that attempt to solve contextual information acquisition and providing data to users [8]. The use and integration of different technologies and the approach to specific learning activities characterize these solutions. However, the complexity of understanding and use of the technology and solutions in the aforementioned works does not allow a wide use of them.

This paper presents the use of CAFCLA [9], a framework aimed at designing, developing and deploying context-aware educational scenarios. Teachers are able to characterize the context where the learning activity will occur through the creation of a world model in which locate data collectors (e.g., sensors), identify and characterize areas of interest (e.g., paintings in a museum), etc. Moreover, the collaboration between students and the customization of the information available is also provided and can be integrated in the activity design. For that, a botanical garden has been chosen as scenario, as it presents the physical and content features to develop a context-aware collaborative learning activity: indoor and outdoor spaces within multiple learning objects (trees and flowers).

The following section describes the background and problem description related to the presented approach. Then, CAFCLA are described: what kinds of activities are covered, how the context of the activity can be defined, who the users are, which activities are implemented by the framework, taking a botanical garden as example scenario. Finally, the conclusions and future work are depicted.

2 Background and Problem Description

A growing interest in educational software, commonly known as e-learning, has appeared over recent years [5]. Mobile devices provide important benefits to education: mobility, communication skills including collection and provision of contextual information, as well as precise location anytime. Mobile Learning is defined as "the processes of coming to know through conversations across multiple contexts amongst people and personal interactive technologies" [10]. This definition implies two important ideas: first of them is that technology can be involved into the learning process; the second idea suggests that mobile learning emphasizes the communication between the involved people and their interaction with the context [11].

Mobile devices allow learners to work outside the classroom to enhance collaborative learning both indoor (e.g., museums) or outdoor (e.g., parks) spaces that present any didactic interest [12]. New mobile devices are equipped with features that facilitate the acquisition of contextual information and location. Contextual information includes any data that can be used to characterize a person, place or object that is considered relevant to the interaction between users, between user and applications or systems, or even between systems and applications [3]. In addition to the relevant information that context provides, it is important to consider other parameters that relevantly affect this type of information, such as identification, time and location [3]. The information exchange taking place between technology and users, in order to contextualize an environment in which learning takes place, and customize the content of the learning activity can be understood as collaboration. Thus, Context-aware Learning must take into account the interactions between people and the different technological components of the system in all its combinations.

Providing contextual information and fostering collaboration between students benefit the learning process [1]. Moreover the combination of Collaborative and Context-aware Learning naturally leads to thinking about ubiquitous learning spaces, characterized by "providing intuitive ways for identifying right collaborators, right contents and right services in the right place at the right time based on learners surrounding context such as where and when the learners are (time and space), what the learning resources and services available for the learners, and who are the learning collaborators that match the learners' needs" [13].

Attending to different technologies there are several trends focused on provide context information to the learning process. A first approach to provide contextual information is "tagging the context". RFID (Radio Frequency IDentification) is the most spread technology [14]. Its use is tedious because teachers have to tag any object they would like to characterize. GPS (Global Positioning System) is the most used technology to provide location in Context-aware Learning [15]. This location system provides a high accuracy level and is currently implemented in a wide range of smart phones and mobile devices. In those cases, the mobile device provides a position to the system. Those solutions are used in different scenarios such as route planning [15]. However, most of those works implement specific applications, but do not propose a general purpose model in which GPS technology is included to facilitate the provision of contextual data. Moreover, teachers cannot decide the activity they would like to implement.

Indoor environments are very common in learning: museums, laboratories or the school are places where activities that require mobility can be developed. Trying to cover this lack, different location systems based on Active RFID [15] or Wi-Fi [16] are used. Both cases the performance of systems is similar: student's position is determined by the access point which is providing coverage in each moment. This type of approach has significant limitations when developing context-aware learning activities: the location accuracy is too poor so teachers are not able to characterize small areas or individual items. This situation presents an important problem when areas where context information is different are close (e.g., two paintings in a museum).

The review of the literature evidences some lacks in the Context-aware Learning systems proposed until now. Even some works try to combine different technologies to cover as much situations as possible [16], teachers are not able to choose different activities to implement or feel free to determine collaborations between students. Moreover, the technology complexity is not transparent to teachers, who are responsible to design the learning activity. Those solutions are focused on the technological part of the problem and do not take into consideration what teachers should do to design the learning process. Next section presents an example of use of CAFCLA that illustrates how this framework takes into consideration the teachers' point of view when designing a context-aware collaborative learning activity.

3   CAFCLA Scenario

CAFCLA (Context-Aware Framework for Collaborative Learning Applications) is a framework focused on the design, development and deployment of collaborative learning applications that make use of contextual information [9]. CAFCLA involves multiple users and characterizes each one according to their role in the design, development, deployment and implementation of a learning activity. Moreover, CAFCLA takes into consideration all aspects surrounding the whole learning process design. These aspects include the objectives or goals that students must reach, the contents of the learning activity, the teaching resources available, the physical or virtual spaces selected or the assessment and activity monitoring. All these aspects do not only involve the teacher, but there is also a technical component that must be undertaken by staff that sometimes do not present an education profile. More specifically, three different roles can be identified in the process of design and development of activities considered in this work.

On first place, teacher is responsible for designing the activity that will be deployed using CAFCLA. Some of the task that teachers carry out includes defining which students participate in the activity, what kind of activity is carried out, which collaborations between students are allowed, which areas and objects of interest are described in the activity, which are the objectives of the activity and which is the data that the system will store to be provided during the activity. Then, developer makes use of all the tools provided by CAFCLA (analysis and design, programming, etc.) to develop the application that students will use, according to the activity designed by teachers. Finally, technician is responsible for deploying the technology infrastructure needed to carry out the activity developed using CAFCLA, following the premises and recommendations set by the framework.

Moreover, student is the participant who finally conducts the activity designed by the teacher. They can access the resources offered by the application through different devices selected for that purpose. In addition, they are able to collaborate between them to achieve the objectives of the activity. Their performance follows the rules set by the teacher at all times.

Once CAFCLA users have been defined, the way in which contextual information is organized is described. According to AmI premises, CAFCLA emphasizes on technological transparency and ease of use for both students and teachers. Contextual information is closely related to the environment where the activity takes place, so any place or item can provide relevant information to be used in the learning process. Thus, teachers are able to describe any place or item relevant to the activity regardless of size and location. In order to better structure contextual information, three description levels have been defined, so that the information can be provided with the granularity required by the activity.

- Scenario: It represents the physical space where the activity will be deployed. To better illustrate the explanation a botanical garden has been chosen to deploy a collaborative learning activity. This scenario consists of an outdoor enclosure where different species of trees, shrubs and flowers grow. Furthermore, in the center of the enclosure there is a greenhouse where multiple flower species grow. In this case the scenario is the botanical garden and it could be divided into two sub-environments: the first one that includes all the study to be performed in the greenhouse (indoor plants), and the second, which would cover the rest of the botanical garden, including all the growth area of outdoor plants.
- Area of interest: Different areas that determine spaces in which a relevant part of the activity will take place. These areas include a physical space where one or several goals of the activity should be reached. The teacher is responsible for identifying, locating and making relevant contextual characterization into them. The areas of interest provide contextual information to the students in the way that the design made by the teacher shows. Continuing the example of the botanical garden, in the external environment three different species of trees grow: pine, oak and poplar. In this case the teacher can create four areas of interest: three individual areas covering spaces where trees grow and a fourth area which is the greenhouse. For each of them, the teacher defines the physical space that it delimits. It also includes a description of each area, based on the design of the activity, that is given to students.
- Object of interest: in the same way that the environments in which the scenario is divided in different areas of interest, within these areas can be included several specific objects that are interesting to the learning activity. Teachers follow the same procedure as in previous cases, since they are responsible for identifying, locating and characterizing these objects. In the example of the botanical garden there may be multiple objects of interest within each of the areas of interest. For example, in the greenhouse grow a wide variety of flowers, and each kind may be an object of interest so teachers are able to identify, place and characterize each one into the greenhouse.

The activities that can be deployed using CAFCLA is another important part of the framework implementation. Different collaborative activities has been evaluated to be integrated by CAFCLA and three of them have been selected: "Treasure Hunting", "Collaborative WebQuest" and "Jigsaw". Different criteria have been taken into account to choose these activities. First of all, these activities can be deployed anywhere and anytime. Secondly, collaboration is possible to be included in all of them. Thirdly, the activities allow teachers to create a learning process that can be monitored and modified at all times. Fourthly, the participants of these activities can be divided into different groups. And finally, all of them can include different routes or physical paths to be followed by the students. However, CAFCLA is an open framework that is able to integrate any other activities that teachers may consider in the future.

Depending on the chosen collaborative activity, the teacher can add the necessary data to complete the learning process. More specifically, the requirements to be considered for each of the activities that are included are described as follows. First activity implemented is a Treasure Hunt. In this activity the teacher can create working groups that are assigned to the corresponding students and determines which devices are used by each students' group. Students that form a group are able to collaborate with each other at all times. Furthermore, the teacher can define different routes that students must follow to uncover clues and collect information. After setting the scenario, the areas of interest and the objects of interest, the teacher defines each route on the map, and indentifies which are the clues given to each group. Routes do not have to be composed of one only path, but may include branches that allow the division of tasks between the different students that are part of the group. The teacher can assign a path or more to a particular group and also may indicate which tracks are key, so that the students are required to complete a milestone to continue receiving information. Finally, the teacher defines a challenge or final question that must be completed or answered with the information received on each track, such as a questionnaire, a document or a presentation to be made.

Collaborative WebQuest is the second activity implemented. The process of defining and describing the scenario, the areas and the objects of interest, the users and the groups of students is common to all the activities. In this activity the teacher is able to design a battery of questions to be answered in each area of interest. Questions can be presented to be answered in a written way or as a test which offers different response options. These questionnaires can be tailored to each users' level and several can be defined for the same area or object of interest. In addition, the teacher can create a final questionnaire to be performed at a particular location (e.g., the classroom) which summarizes all questions regarding the questionnaires that have been made in the activity. Likewise, the teacher has the capacity to define different working groups formed by the number of students it may consider, assigning devices that each group or student uses. Likewise, the teacher indicates the questionnaires that each group must respond. The onset of the activity is performed for each group in the particular zone determined by the teacher through a first task in explanation of the activity.

The last implemented activity is a Jigsaw. As in the previous activities, the process of defining and describing the scenario, the areas and the objects of interest, the users and the groups of students is the first step to be completed. In this activity the teacher divides the activity into different subjects to study individually and then in different groups. First, the teacher assigns each student a specific topic in which will be "expert". Later, two different types of groups are formed: on the one hand by students who have been assigned with different topics (each group will consist of an expert in each of the different themes described) and on the other hand by "expert" students in the same topic. Similarly, the teacher determines which of the areas of interest that have been created belongs to each of the assigned topics. For the proper operation of the activity, the teacher indicates what documents will be generated at each of its stages: single phase, experts phase (collaboration among students working under the same topic) and group phase (collaboration among students working on different topics). The final result of the activity (e.g., a presentation) should be exposed by the group's leader, a role that is assigned by the teacher.

4 Conclusions and Future Work

The use of Information and Communication Technology in the different areas has increased in recent years thanks to the emergence in society of mobile devices, easy access to currently existing technology and the many features they present, such as communication protocols and context-aware technologies. However, it is difficult to develop applications to squeeze all the potential offered by technology, especially when the main objective is the development of technological applications that are transparent to users.

CAFCLA has been designed with the objective to design and develop a set of tools that provide a basis for designing, developing and implementing context-aware collaborative learning activities. CAFCLA is a framework that integrates different context-aware technologies, such as Real Time Locating Systems, and several communication protocols that abstract educators and developers of context-aware collaborative learning activities from the complexity of the use of different technologies simultaneously. In this case, CAFCLA focuses on provide a set of tools and methods to teachers, developers and technical staff in order to easily design, develop and deploy this type of learning activities.

The development of the framework has been approached from the teacher's point of view. Thus, the design of different kind of collaborative activities has been implemented by CAFCLA focusing in the teacher's work and how they can include information and collaboration between students.

Future work includes the design, development and deployment of a real use case where all the features of CAFCLA are implemented. This work will be developed by different teachers and developers in order to compare the results reached by all of them and evaluate the framework in a real scenario.

Acknowledgments. This project has been supported by the Spanish Ministry of Science and Innovation (Subprograma Torres Quevedo).

References

1. García, Ó., Tapia, D.I., Alonso, R.S., Rodríguez, S., Corchado, J.M.: Ambient intelligence and collaborative e-learning: a new definition model. Journal of Ambient Intelligence and Humanized Computing, 1–9 (2011)
2. Jorrín-Abellán, I.M., Stake, R.E.: Does Ubiquitous Learning Call for Ubiquitous Forms of Formal Evaluation?: An Evaluand oriented Responsive Evaluation Model. Ubiquitous Learning: An International Journal (2009)
3. Dey, A.K.: Understanding and Using Context. Personal and Ubiquitous Computing 5(1), 4–7 (2001)
4. Scardamalia, M., Bereiter, C., McLean, R.S., Swallow, J., Woodruff, E.: Computer-Supported Intentional Learning Environments. Journal of Educational Computing Research 5(1), 51–68 (1989)
5. Gómez-Sánchez, E., Bote-Lorenzo, M.L., Jorrín-Abellán, I.M., Vega-Gorgojo, G., Asensio-Pérez, J.I., Dimitriadis, Y.A.: Conceptual framework for design, technological support and evaluation of collaborative learning. International Journal of Engineering Education 25(3), 557–568 (2009)
6. Laine, T.H., Joy, M.S.: Survey on Context-Aware Pervasive Learning Environments. International Journal of Interactive Mobile Technologies 3(1), 70–76 (2009)
7. Bruce, B.C.: Ubiquitous learning, ubiquitous computing, and lived experience. In: Cope, W., Kalantzis, M. (eds.) Ubiquitous Learning, pp. 21–30. University of Illinois Press, Champaign (2008)
8. Chen, T.-S., Yu, G.-J., Chen, H.-J.: A framework of mobile context management for supporting context-aware environments in mobile ad hoc networks. In: Proceedings of the 2007 International Conference on Wireless Communications and Mobile Computing, pp. 647–652 (2007)
9. García, Ó., Alonso, R.S., Tapia, D.I., García, E., De la Prieta, F., de Luis, A.: CAFCLA: A conceptual framework to develop collaborative context-aware learning activities. In: Uden, L., Corchado, E.S., De Paz, J.F., De la Prieta, F. (eds.) Workshop on LTEC 2012. AISC, vol. 173, pp. 11–21. Springer, Heidelberg (2012)
10. Sharples, M., Taylor, J., Vavoula, G.: A Theory of Learning for the Mobile Age. In: Bachmair, B. (ed.) Medienbildung in neuen Kulturräumen, pp. 87–99. VS Verlag für Sozialwissenschaften, Wiesbaden (2010)
11. Glahn, C., Börner, D., Specht, M.: Mobile informal learning. In: Brown, E. (ed.) Education in the Wild: Contextual and Location-based Mobile Learning in Action, Nottingham. A report from the STELLAR Alpine Rendez-Vous Workshop series, pp. 28–31 (2010)
12. Neyem, A., Ochoa, S.F., Pino, J.A., Guerrero, L.A.: Sharing information resources in mobile ad-hoc networks. In: Fukś, H., Lukosch, S., Salgado, A.C. (eds.) CRIWG 2005. LNCS, vol. 3706, pp. 351–358. Springer, Heidelberg (2005)

13. Hwang, G.-J., Yang, T.-C., Tsai, C.-C., Yang, S.J.H.: A context-aware ubiquitous learning environment for conducting complex science experiments. Computers & Education 53(2), 402–413 (2009)
14. Blöckner, M., Danti, S., Forrai, J., Broll, G., De Luca, A.: Please touch the exhibits!: using NFC-based interaction for exploring a museum. In: Proceedings of the 11th International Conference on Human-Computer Interaction with Mobile Devices and Services, MobileHCI 2009, vol. 71, pp. 1–2 (2009)
15. Driver, C., Clarke, S.: An application framework for mobile, context-aware trails. Pervasive and Mobile Computing 4(5), 719–736 (2008)
16. Martín, S., Peire, J., Castro, M.: M2Learn: Towards a homogeneous vision of advanced mobile learning development. In: 2010 IEEE Education Engineering (EDUCON), pp. 569–574 (2010)

Learning to Be an Entrepreneur: Evidence from a Technology Enhanced Learning Tool

Gianfranco Giulioni, Edgardo Bucciarelli, Marcello Silvestri, and Paola D'Orazio

Abstract. In this paper we present a web-based software model in which the user manages a virtual enterprise. By using the software the user experiences a number of basic economic principles such as the leverage effect, the effect of depreciation on the firm return rate and the key role of the equity base in a framework characterized by risk and uncertainty. The model has been used in an undergraduate class at the Faculty of Management Sciences at the University of Chieti - Pescara. Students have acknowledged the usefulness of this tool in order to assess the theoretical principles learned during the lectures. We then asked a different subject pool, i.e. entrepreneurs, to manage the same GUI. A comparative analysis of the results obtained in the different sessions showed that our software is a good and effective tool for learning purposes.

1 Introduction

In the last decade technology-enhanced learning has gradually been drawing considerable attention [2]. In particular, the widespread deployment of computing devices has played a key role in opening new frontiers for the development of learning activities.

Among them, *mobile learning games* (for a survey, see [11]) are of particular importance. In the literature it is indeed widely acknowledged that they are useful for learning and teaching purposes as they provide "affective" and cognitive learning outcomes [6, 12].

Within this orientation we provide an economic game based on Hyman P. Minsky theory. By means of a web application we shape a system wherein subjects could

Gianfranco Giulioni · Edgardo Bucciarelli · Marcello Silvestri · Paola D'Orazio
Department of Economical-Statistical and Philosophical-Educational Sciences, Viale Pindaro 42 - 65127, Pescara, Italy
e-mail: {g.giulioni,e.bucciarelli,paola.dorazio}@unich.it,
    marcello.silvestri@yahoo.it

P. Vittorini et al. (Eds.): *2nd International Workshop on Evidence-Based TEL*, AISC 218, pp. 67–74.
DOI: 10.1007/978-3-319-00554-6_9   © Springer International Publishing Switzerland 2013

be able to perform their game regardless of their geographic location. Our software could be indeed run just pointing the browser to an URL and inputing username and password. Moreover, by using the Java language the software is able to gather and store data on a remote database.

As regarding informal learning, the web-based software model described in the paper is based on an innovative experimental economics technique (for a survey see, [8], [5]) which allows to explore a number of economic and financial aspects of the entrepreneurial management.

The paper is organized as follows. In section 2 we describe the theoretical framework and how it was implemented in the web-based software. Section 3 shows the evidence based on students and professional performances. Section 4 concludes.

2 A Technology Enhanced Learning Tool

Starting from a theoretical model rooted in the Hyman P. Minsky's theory of the firm [9] we built a virtual dynamic corporate environment where subjects behave as entrepreneurs and/or managers by taking economic and financial decisions which will determine their utility (i.e. the final score). In the next section we briefly summarize the virtual environment theoretical (subsection 2.1) and operating (subsection 2.2) principles.

2.1 Theoretical Background

The *technology enhanced learning tool* we built intends to be a small scale abstraction of the real world and it is meant to capture the essence of the Hyman P. Minsky economic thought. The theoretical aspects of the present work are summarized hereafter. For a more detailed presentation of the theoretical aspects see Giulioni et al. (2011) [7].

According to the economic theory which underlies our software, the process of decision making (i.e. production and balance sheet structure decisions) is characterized by *uncertainty*. The balance sheet consists of a set of assets owned or controlled by firms and a set of liabilities used to finance those assets. Assets and liabilities involve cash-flows and payment commitments in a delayed future. Cash-inflow depends on sales of the produced items which in turn depend on the level of inputs employed in the production process. In our simplified environment the value of production inputs is equal to the sum of the firm balance sheet assets entries. Firms fulfill commitments on liabilities (interest and dividends payments) using the cash-inflow that will be realized in a delayed term. Each firm must decide the "proper mix" of assets and liabilities and how to manage their own capital structure according to their *subjective perception* of the demand volatility for the produced item. This can be defined as a "traditional" portfolio choice which is invariably linked to payment obligations undertaken in the past and to the uncertainty that affects future prospects, which in turn must validate such obligations.

The scenario subjects have to interact with is described as follows. Subjects are asked to manage a virtual operating firm. Accounting for some preliminary information (a forecast on the future level of demand and past realization of the same variable which are continuously updated), subjects are asked to set the level of their physical capital which in turn determines the optimal scale of production, i.e. the level of production which minimizes production costs. Once this choice has been made the software reveals the level of demand. The maximum profit is realized if the optimal production scale is equal to the received demand. The difference between production revenues and costs decreases as long as the gap between the optimal production scale and the level of demand increases. For high levels of the gap, productions costs become progressively higher than revenues and the firm suffers an increasing loss.

Beyond this production revenue/cost structure, financial burdens (interest and dividends) have to be accounted for. The economic result of the firm activity is indeed obtained by subtracting production and financial costs from the sales influx and it can be either positive or negative. Concerning financial costs, we set the cost of the equity base higher than the interest on debt following Myers and Majluf [10]. In this way subjects have to take a basic *speculative decision*: a high level of the equity base ensures entrepreneurs against prospective losses but at the same time it reduces the rate of return of the entrepreneurial activity. In our settings, a firm can have serious consequences after a number of consecutive negative economic results: the equity base may reach low values so that a further loss could lead the equity base to negative levels. In this case the software requires subjects to activate a bailout procedure which has a cost (i.e. it implies a penalty).

The experimental subjects' performance is evaluated according to a score computed by the software as the average return on investment minus the bailouts penalties.

2.2 The Web-Based Software Model

In this subsection we describe the functioning of our web-based software model. Figure 1 shows the Graphical User Interface (GUI) related to the decision process described above[1].

The GUI is composed of a control panel and three charts.

The control panel is divided into three specific sections. The top one displays information on the figures supplied by the forecasting service, the production choices, the market demand realizations and the errors made in guessing such levels concerning the time of the choice and three past periods. In the second section labeled as the "patrimonial and economic accounting" the user monitors the balance sheet items and the debt ration looking at the stock line. The economic result is showed under the label "profit" together with its maximum achievable value, the Return on

---

[1] For those interested in experiencing a trial of the software, point the browser to
http://www.dmqte.unich.it/users/giulioni/
demo_en/experiment.html

Investment, the average return on investment, the bailouts and the score in the flows line. The third section is composed by the fields where subjects have to input their production and financing choices according. It deserves a close examination since it focuses on the operational part of subjects and on their speculative decisions.

User choices. Taking into account the information from the charts and from the other sections of the control panel subjects have to set first the level of production "guessing" on the future level of demand. Once subjects input the figure in the production choice field and press the return key, the software updates the balance sheet structure: the equity base is not affected so that debt and assets adjust according to changes in the production capacity. The Graphical User Interface reveals the level of demand, computes the economic result and activates the appropriate choice field. Now, the user is asked to management the balance sheet structure of the firm.

An analysis of the *economic result* is helpful for the understanding of the functioning of this part of the software. As mentioned above, the economic result can be either positive (profit) or negative (loss).

When a *profit* is realized, it can be used to refund the bank. In this case the key decision is on the amount of profit to be used in order to reduce the debt. To this aim, the "decrease debt" field is enabled and the user is asked to input a value. Assets are not affected by this choice so that the reduction of debt implies an equal increase in the equity base. Subsequently subjects have to input a value in the "decrease equity" field. As mentioned above, we assume that the reward for the shareholders is more expensive than the interest rate charged by the bank so that subjects can increase the leverage ratio of the firm to the aim of obtaining a higher rate of return. Once this choice is made, the prompt jumps to the next period production choice.

If a firm suffers a *loss*, the flow of event is different and users can face two different situations which in turn depend on the amount of equity they hold (i.e.

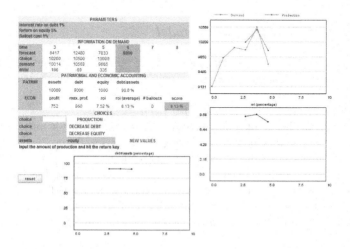

Fig. 1 The Graphical User Interface of the web-based software model

whether it is enough to face the loss). In both cases the assets and the equity base are reduced by the loss. When equity base is enough to face the loss, the firm survives. In this case the software skips the "decrease debt" input field, and it activates the "decrease equity" field. When the loss is higher than the equity base, the decisions on debt and equity reduction are both skipped and the prompt jumps to the bailout section before going to the next period production choice.

Implementing the Evidence Based Learning. The forecasting service as well as demand trends are built perturbing a regular benchmark pattern with random noises. The standardized percentage deviation of perturbations follow a probability distribution chosen by the researcher so that both the accuracy of forecasts and the volatility of demand can be exogenously tuned. Standard deviations of demand and forecast perturbations are crucial in determining the degree of riskiness in the virtual corporate environment. By changing these parameters, the instructor can make users to experience situations ranging from easily manageable to highly turbulent. The user can learn how to modify the strategy s/he uses to improve her/his performances by repeating the exercise with unchanged settings. Furthermore, a valuable educative task can be achieved by comparing the best strategies found for different operational contexts.

Some evidence on the learning process implemented by using the software are reported in the following section.

3   Evidence

We report hereafter some results of exercises we have performed. The first exercise is characterized by a very reliable forecasting service: the value supplied is always equal to the level of demand which will be received by the firm. At first glance, the exercise seems to be trivial because the experimental subjects know in advance the level of demand and consequently they adapt the production capacity precisely. This apparently trivial exercise has a hidden mechanism: the leverage effect. As discussed above, in the proposed setting debt is cheaper (1% interest rate) than equity (the entrepreneur must pay 5% to shareholders). Equity is a buffer against losses, but with perfect forecast one cannot suffer losses. Taking all this in mind, the best performing strategy is a 100% debt liability structure. In this context, experimental subjects prove to be fast in learning to use the supplied forecast. However, heterogeneity in learning the leverage effect is present. In the exercises we have performed, the experimental subjects were allowed to repeat the exercise as many times as they wanted with the aim of improving their score.[2]

The results obtained from our activity concerning the perfect forecast context are reported in tables 1 and 2 which reports the final scores of the subjects.

[2] For a survey of the literature on the learning to forecast "ability" of economic agents and in order to gain more insights on this topic see [3], [4], [1].

Table 1 shows results from undergraduate students at the Faculty of Management Sciences at the University of Chieti - Pescara. The same experiment was performed with selected entrepreneurs whose result are reported in figure 2. The first column of the tables reports the subject's identification number.[3] The following columns inform on the scores achieved by the subject in consecutive trials. The black slots identify the achievement of the maximum score. It is important to highlight that the score goes from 5% (when equity cover 100% of liabilities) to 9% (when debt is 100%). It can be observed that a significant share of the subjects (both students and entrepreneurs) have not yet taken advantage of the leverage effect when they left the computers room.

Table 1 Results from a class of students at Management Sciences Faculty

01	6.97	7.44	8.69	8.93				
03	8.33	8.34	**9**					
06	7.84	8.21	8.02	8.49	8.07	8.63	8.6	8.4 **9**
07	8.24	8.51	8.69	8.83				
08	8.45	8.77	8.82	8.9	**8.99**			
09	8.9	8.97	**9**					
10	7.34	7.77	6.79	8.59	8.11			
11	8.69	8.82	8.96	8.98	**9**			
12	6.75	8.97	8.94					
13	8.45	7.96	8.59					
14	8.8	**9**		8.97	8.88	**9**		
15	8.35	8.65	8.81	8.15	8.77	8.79		
16	8.96	**9**		**9**				
17	7.68	8.85	**9**		**9**			
18	8.38	8.57	8.85	8.93	**9**			
19	7.05	8.78	**9**					
20	7.58	8.98	**9**					
21	8.45	8.88	8.92	8.92				
22	7.17	7.6	8.95					
23	8.74	8.78	8.69	8.82				

The second exercise was carried out accounting for a risky environment. We asked the experimental subjects to take part to three experimental sessions (labeled as B1a, B1b and B1c) which are characterized by a decreasing "accuracy" of forecasts and a low level of the volatility of demand. In this case, the aim is to investigate the switching behavior between a *forward looking* and a *backward looking* conduct. Intuition suggests that when forecasts are precise, agents should rely on them. In the case of imprecise forecast, good "guesses" on future demand could still be obtained by looking at the past if the demand time series is smooth (i.e. it has a low volatility). We present hereafter the results of linear regressions performed in order to investigate the experimental subjects' switching conduct. Since experiments with the entrepreneurs' subject pool is still at an early stage, we report hereafter only students' results. The estimated equation is:

Table 2 Results from entrepreneurs selected from the Pescara entrepreneurs union

01	8.35	8.81	**9**		
02	7.71	7.04	6.76	7.4	8.5
03	8.66	8.54	8.84		
04	6.44	8.66	8.82	**9**	
05	8.09	8.26	8.35		
06	8.82	8.98	**9**		
07	8.5	8.93	**9**		

$$y_{j,e,t} = \alpha_{j,e} + \beta_{j,e} y^{*}_{j,e,t-1} + \phi_{j,e} \tilde{y}_{j,e,t} + \varepsilon_{j,e,t} \qquad (1)$$

where $y_{j,e,t}$ is the production capacity chosen by subject j in exercise $e \in \{B1a, B1b, B1c\}$ at time t; $y^{*}_{j,e,t-1}$ and $\tilde{y}_{j,e,t}$ are respectively the past demand realization and the current level of forecast and $\varepsilon_{j,e,t}$ is the error term.

We ran a regression for each subject in each of the three experiments. The results show that subjects in general sets the optimal production capacity as a weighted average of the previous period demand and the forecast. Surprisingly enough, the

---

[3] In tables 1 and 3, the identification numbers 2, 4 and 5 are missing because the students associated with them do now attended the session.

sum of the two slopes parameters is aprroximately equal to one for a large number of the subjects. Taking this in mind, we report in table 3 the estimates for the $\beta_{j,e}$s. The comparison of the coefficients for each subject in the three experiments shows how, except in a number of occasions, with rising standard deviation of forecast subjects set the production choice closer and closer to the past demand realization (β progressively increases going trough $B1a$, $B1b$ and $B1c$). Subject 10 and subject 13 represents two exception because they heavily rely on forecasting service when chosing the production capacity in all the three experiments.[4] Furthermore, subjects 14, 16, 19, 20 and 22 are those who set the production choice according to both backward and forward looking. These early results show a degree of heterogeneity agents' learning how to manage risky and uncetain environments.

Table 3 Estimates of the $\beta_{j,e}$ parameter present in equation (1) by using OLS regression. The subjects pool is the class of students at Management Sciences Faculty.

e	j = 1	j = 3	j = 6	j = 7	j = 8	j = 9	j = 10	j = 11	j = 12	j = 13
B1a	0.182*	− 0.156**	0.600***	0.389***	0.011	− 0.237*	0.002	0.195***	0.018	0.228***
	(0.090)	(0.075)	(0.060)	(0.084)	(0.044)	(0.134)	(0.016)	(0.052)	(0.032)	(0.077)
B1b	0.240***	0.469***	0.364***	0.427***	0.115**	0.896***	0.089	0.423***	0.441***	− 0.071
	(0.067)	(0.058)	(0.063)	(0.057)	(0.065)	(0.045)	(0.092)	(0.081)	(0.081)	(0.056)
B1c	0.999***	0.816***	0.726***	0.791***	0.807***	0.769***	− 0.070	0.767***	0.816***	0.079
	(0.075)	(0.055)	(0.032)	(0.055)	(0.098)	(0.083)	(0.044)	(0.045)	(0.067)	(0.071)

e	j = 14	j = 15	j = 16	j = 17	j = 18	j = 19	j = 20	j = 21	j = 22	j = 23
B1a	0.533***	0.355***	0.120**	0.350***	− 0.107*	− 0.006	-0.033	0.910***	0.298***	0.050
	(0.052)	(0.060)	(0.046)	(0.052)	(0.054)	(0.040)	(0.102)	(0.071)	(0.062)	(0.055)
B1b	0.295***	0.613***	0.389***	0.596***	0.239*	0.265***	0.395***	0.958***	0.386***	0.491***
	(0.042)	(0.064)	(0.070)	(0.067)	(0.011)	(0.140)	(0.088)	(0.067)	(0.039)	(0.064)
B1c	0.638***	0.758***	0.578***	0.779***	0.801***	0.345***	0.586***	1.009***	0.644***	0.838***
	(0.078)	(0.055)	(0.041)	(0.068)	(0.056)	(0.109)	(0.049)	(0.039)	(0.045)	(0.049)

Notes: ∗ ∗ ∗, ∗∗, ∗, denote significance at $p = 0.01$, 0.05, and 0.10, respectively (twotailed tests). Standard errors are in parentheses.

4 Conclusions

This paper describes a web-based software model which is a small scale abstraction of the real world characterized by risk and uncertainty.

The results of our experiments with two different subject pools, i.e. entrepreneurs and students, showed that the model could offer a suitable framework for the exploration of agents' behaviors in a *stochastic* and *dynamic* context. Our software allows in particular to observe how agents learn to forecast and manage the capital structure of their firms.

We made use of Experimental Economics which has been drawing a growing attention among researchers since it allows for the observation of individuals' behaviors. In particular our software showed that the experimental method could be

---

[4] One can imagine that they they have very low scores especially in experiment $B1c$ because of the large number of bailouts they activate.

a useful tool to improve learning. We thus promote a combination between the scientific methodology of experimental economics and technology enhanced learning tools. The former provides a theoretical framework in which subjects could learn while the latter represent the support for the learning activity as a whole.

We maintain that the coupling of informal and theoretical learning is a fruitful experience for students. On the other hand, the use of technology enhanced learning tools entails soft skill training which can be particularly suitable for education management, recruitment and selection as well as for the assessment of managers' performances and their attitude in achieving business goals.

References

1. Anufriev, M., Hommes, C.: Evolutionary selection of individual expectations and aggregate outcomes in asset pricing experiments. American Economic Journal: Microeconomics 4(4), 35–64 (2012),
   http://ideas.repec.org/a/aea/aejmic/v4y2012i4p35-64.html
2. Balacheff, N., Ludvigsen, S., de Jong, T., Lazonder, A., Barnes, S.: Technology Enhanced Learning. Springer, Berlin (2009)
3. Bao, T., Duffy, J., Hommes, C.: Learning, forecasting and optimizing: an experimental study. CeNDEF Working Papers 11-08, Universiteit van Amsterdam, Center for Nonlinear Dynamics in Economics and Finance (2008),
   http://EconPapers.repec.org/RePEc:ams:ndfwpp:11-08
4. Bao, T., Hommes, C., Sonnemans, J., Tuinstra, J.: Individual expectations, limited rationality and aggregate outcomes. Journal of Economic Dynamics and Control 36(8), 1101–1120 (2012), http://ideas.repec.org/a/eee/dyncon/v36y2012i8p1101-1120.html
5. Duffy, J.: Macroeconomics: A survey of laboratory research. Working Papers 334, University of Pittsburgh, Department of Economics (2008),
   http://ideas.repec.org/p/pit/wpaper/334.html
6. Facer, K., Joiner, R., Stanton, D., Reidz, J., Hullz, R., Kirk, D.: Savannah: Mobile gaming and learning? Journal of Computer Assisted Learning 20, 399–409 (2004)
7. Giulioni, G., Bucciarelli, E., Silvestri, M.: A model implementation to investigate firms financial decisions (2011),
   www.dmqte.unich.it/users/giulioni/model_description.pdf
8. Kagel, J.H., Roth, A.: Handbook of Experimental Economics. Princeton University Press, Princeton (1995)
9. Minsky, H.P.: Induced Investment and Business Cycle. Edward Elgar Publishing, Northampton (2004)
10. Myers, S.C., Majluf, N.S.: Corporate financing and investment decisions when firms have information that investors do not have. Journal of Financial Economic 13, 187–221 (1984)
11. Schmitz, B., Klemke, R., Specht, M.: Effects of mobile gaming patterns on learning outcomes: a literature review. International Journal of Technology Enhanced Learning 4, 345–358 (2012)
12. Spikol, D., Milrad, M.: Physical actiivties and playful learning using mobile games. Research and Practice in Technology Enhanced Learning 4, 345–358 (2012)

USALSIM: Learning and Professional Practicing in a 3D Virtual World

Emiliana Pizarro Lucas, Juan Cruz Benito, and Oscar Gil Gonzalo

Abstract. USALSIM was developed by the University of Salamanca (USAL) as a response to the policy changes of learning and work placements in the new European Space for Higher Education. USALSIM makes it possible to face the increase of the number of students who will participate in the different university work placement programs and the increasing number of companies and institutions necessary to host and train theses students. This project used a 3D virtual environment (a work placement simulator) that allows developing a virtual representation of different work environments. Representing a professional work space such as a laboratory, for example, the student can simulate common tasks through active learning. This virtual world is focused on a constructive pedagogy, where students are directly involved in their formative development, establishing professional relationships, developing transversal and technical competencies and evaluating their knowledge. USALSIM is a funded project by the Spanish Ministry of Education within the Program of Integral Attention and Employability of University Students (CAIE059).

Keywords: University of Salamanca, USAL, Service of Professional Insertion, Practices and Employment, SIPPE; OpenSimulator, OpenSim, Metaverse, Education, E-learning, Work Space, Practices, Virtual World, Moodle, Sloodle.

1 Introduction

There are different definitions of what constitutes a virtual world and what its primary characteristics are. According to [1][2], a virtual world environment is " *[a]computer-generated display that allows or compels the user (or users) to have a sense of being present in an environment other than the one they are actually in, and to interact with that environment*".

Emiliana Pizarro Lucas · Juan Cruz Benito · Oscar Gil Gonzalo
Service of Professional Insertion, Practices and Employment, University of Salamanca,
Patio de Escuelas 1, 37008, Salamanca, Spain
e-mail: {mili,juancb,oscar.gil}@usal.es

P. Vittorini et al. (Eds.): *2nd International Workshop on Evidence-Based TEL*, AISC 218, pp. 75–82.
DOI: 10.1007/978-3-319-00554-6_10   © Springer International Publishing Switzerland 2013

Over the last decade there have been several studies regarding the advantages of multi-user virtual environments for teaching, due to their relationship with the concept of *experiential learning*. According to Kolb's theory of experiential learning [3], it is possible for the learner to gain experience and deeper knowledge by experiencing the four-stage learning cycle, which consists of: Concrete Experience, Reflective Observation, Abstract Conceptualization, and Active Experimentation. Other experts such as [4][5] assert that the potential of 3D virtual worlds is rooted in the possibility of offering more student-centered learning processes, which require students to be more committed by relying on experiential activities.

This and other theories were used as the starting point for USALSIM. This paper aims to review the basic concepts used in the development of the 3D environment of USALSIM, the technologies employed, and the conclusions reached. Section 2 presents the needs which gave rise to the project. Section 3 describes the basic technologies used and the defined functional structure created for the simulation environment created of the project. Finally, section 4 presents the various lessons learned and posits future research.

2   Background

The USALSIM project aims to respond to the needs associated with professional training and work placement within the current context of university studies, and to resolve the difficulties that arise from coordinating these placements, such as finding the number of companies needed to develop the work placement program, and coordinating school and work schedules. By using a "training simulator" that develops a virtual 3D world, we hope to expand existing possibilities by representing occupations and professional relationships that replicate daily activities, so that students can enhance their own training.

In the case of professional work placements that are developed outside the educational center, USALSIM permits a mixed model that combines the training received in a virtual space with those developed in a real work space. Training received through a "training simulator" can be done prior to or during the actual work placement, and both are supervised equally by a mentor in the company and an academic advisor, in addition to the collaboration of a professional guidance counselor. This enables students to use virtual reality to learn tasks that will be physically carried out both in a work placement environment and in the future, throughout the course of their professional career. This mixed model lends itself to variations, allowing the professional activities performed in virtual reality to be developed prior to or during the period of work placement, to complement the hours worked in the company, to serve as an evaluation tool for the academic advisor and mentor, to complement the theoretical and practical syllabus, to assist the student's entrance into the workforce, to represent a new experience in social interaction, etc.

From an education perspective, USALSIM will serve to both assist and complement specific activities that are occasionally unfeasible to be carried out within the University. The project can provide spaces that are not always available in real life, making it viable for students to prepare themselves or train before engaging in specific processes or activities, or provide environments that can be used for meetings, work exhibits, conferences, etc.

3   Developing the USALSIM Virtual Word

USALSIM has developed a virtual world in 3D that represents the most common work spaces of five professional areas (Pharmacy, Law, Biology, Biotechnology, Chemistry, and Humanities). The *OpenSimulator [6]* application server was used to develop the virtual environment. This platform software is used to create multi-user and multi-platform virtual worlds in 3D. It is an open code and is supported by a wide user community.

With *OpenSimulator* (*OpenSim*) the 3D virtual world is organized according to regions, or patches of land, which could be an island, a country, or a mega-region, etc. In the case of the USALSIM project, the organizational architecture in the initial phase opted for space in the form of islands. Five regions (islands) were built, each one corresponding to a title or branch of knowledge, and all joined to a sixth island representing the Professional Integration Service, Training and Employment of the University of Salamanca, the department where the project is managed.

For each of these regions or island, a set of training tasks, both educational and professional, were developed with very specific objectives. These trainings were designed with the cooperation of educators and professionals currently working in the respective fields. The content of the training, once it was designed, was adapted to the technical and functional requirements of the simulator. Three types of training were identified:

- *Informative training,* which provides students with the theoretical base needed to perform other tasks at a later date.
- *Procedural training,* which prepares or trains the student to deal with procedures or activities during the training period.
- *Immersion training,* which accurately imitates the reality the student may face in a working environment, and prepares the student to deal with life-like situations in the simulator.

For the development of the USALSIM virtual world, a functional architecture was defined over which the different trainings for each "island of knowledge" were created. This architecture is based on three standard practice models, defined by a set of entries, exits, objects and connections among objects. Each training model defines the effect derived from the user's action with one or more of the objects that populate the space, bearing in mind a series of preconditions and post conditions. On that basis, the training models are distinguished as follows:

• *Distributed training model: According to [7] "a distributed system is a collection of independent computers that give the user the impression of constituting a coherent system". In this model, the "Distributed System" is not a collection of computers, but of three-dimensional objects (laboratory equipment, etc.) that execute autonomously a series of tasks and interact among themselves in a distributed way to develop the training performed by the user. Each 3D object (peer) communicates to the others the result of the internal operation produced by the user's action, or by the action of another object, such that each object is consequently able to react when the user interacts with it. In this horizontal type of communication, all of the peers are alert to any communication from the others. In this case, no single object coordinates or becomes the center of communication. This allows all objects in a defined area to be continuously updated so that a high volume of interaction does not produce a bottleneck or inactivity due to the lack of information in any of the objects of the defined area. This distributed architecture of 3D objects is based on a graph structure, that is, on a set of objects (vertexes) joined by links (artists) representing binary relationships among the set of elements.*

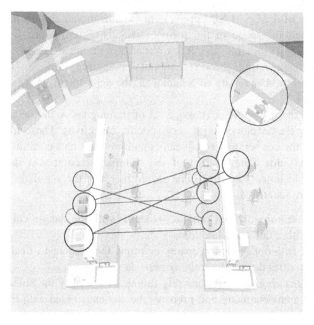

Fig. 1 Visualization of the object relationships in a training set

This training model presents a potential problem: all objects contain most or all of the logic of the training, meaning that an object with a high level of interaction can function at a slower processing speed. On the other hand this type of architecture allows the training to continue operating even if a node fails, as long as it is not critical (is part of a required step).

- *Star training model*: in this topology a single object coordinates the various relationships among the other training elements. That is, a specific 3D object acts as coordinator for the remaining 3D objects in the space. All communication passes through this object, and it alone decides what to do with the incoming information. This topology is applied in training sessions where the collaborative process is strictly defined and has a determined sequencing, so that the number of relationships among objects is small. In this way, there are satellite objects with a low execution load, and an object that contains all the logic of the training and makes decisions regarding the global process.
- *Role-playing training model*: training content and resolutions can only be reached by interacting with other users (avatars). For instance, the training of Law includes exercises related to judicial processes that would have to be simulated in a classroom environment since real courthouses are not available to conduct this activity. Thus, virtual spaces representing a courtroom were developed in order to create an immersion style training as close as possible to a real-life environment, even including the appropriate attire. These trainings require interaction among various users playing the different roles required in the judicial process.

Fig. 2 View of virtual courtroom training

- In this training model there is no definite execution prescribed by the system for a set of 3D objects. Instead, the users are provided with a set of interactive tools to develop the training. The tools that they are given to complete the training are basically communication tools such as text or voice. Additionally, they may have access to interactive screens, video players, screens for presentations, etc.

From the beginning of this project, one of the specific requirements established was the ability to connect the virtual world with *Moodle* [8] technology, as it is the virtual learning platform used by USAL. This platform is used as a distance learning tool and as a support tool for in-class learning. It has been adapted to the specific needs of this university. The professors use the platform to make educational materials and complementary references available to the students, in addition to monitoring and evaluating their performance through tests and questionnaires, or offering consulting and tutorial mechanisms (forums, chatrooms, videoconferences…). Given that USALSIM is an education tool, there are several advantages to integrating these complementary platforms. The *Sloodle* [9][10][11] project was used to integrate the two platforms. It is an open code project that permits the integration of multi-user virtual environments such as *OpenSim* and the *Moodle* system.

The integration of *Moodle* with USALSIM makes it possible to create within a virtual many things, including: presentations housed in the university's e-learning platform, link user profiles from both platforms to share the data of registered users, the possibility of administering tests to users, engage in chats, etc. With regard to the tests or questionnaires, the user may request to perform them in a virtual environment. In this case, the test is downloaded from the server and visualized through a series of text boxes, and the user selects the correct option by pushing a button; once the test is completed, the user is informed of the result and the Moodle platform saves all data related to the text (duration, score, etc.) just as if the test had been taken on the e-learning platform.

4 Tests and Evaluation

In order to validate the system, we proposed several professional situations, distributed in the different jobs corresponding to the knowledge areas discussed above. In this stage of our testing, we established a group of 139 users (students and teachers or professionals) to perform the different tasks. At the end, 86 of these people filled in diverse questionnaires where they explained their views and opinions about the system, its design, use, utility and possible application in their field of study.

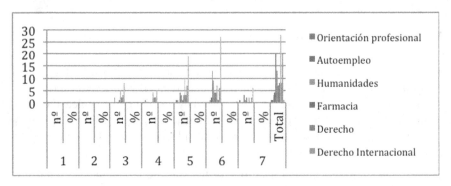

Fig. 3 Students Evaluation

We show an example of this analysis in the attached graph, where 67 students assessed their satisfaction with the simulator using a score ranging between 1 and 7 on a Licket scale.

5 Conclusions

USALSIM was born as a local project of the University of Salamanca, hence the need to develop a cost-controlled proprietary software platform. This is where the use of free software has played an important role. Other virtual world platforms could have been adopted, such as *Second Life*, with stable versions and advanced technical support. On the other hand was the issue of maintenance costs, as each island created must be purchased and maintained through fee payments during the year (Premium service). Moreover, according to the terms and conditions of the software agreement, any 3D construction in the platform belongs to the owners of *Second Life*. To all of this was the added issue that real and sensitive user data would not be under the control of the University, but in the hands of a third party. For these reasons, the University decided to go with an open code such as *Open-Sim*, which allows user data (and complete data bases) to be stored in the local servers of the University. Additionally, there are no added costs related to the construction of each virtual 3D space, no Premium Services, and the access code is free and universal, allowing for its modification or adaptation to the specific needs of the University.

After the development of the virtual space, a number of tests were performed with real users. There were 150 participants, students from different branches of knowledge, as well as professors, counselors, personnel associated with participating entities and companies, and other University personnel. In this test phase, users performed their assigned trainings and provided their opinion regarding the system through the use of questionnaires. The questions asked them to rate different aspects such as ease of installing the software needed to connect to the 3D virtual environment (connection client), ease of configuration, ease of moving about the virtual environment, ease of performing the training, in addition to suggestions for improvements. The evaluations from the student users were very positive, with the vast majority approving the use of the simulator and considering the use of this type of technology both a convenient and advantageous way of complementing their studies. In addition to the students, a similar evaluation of this tool was given to educators and personnel from the different participating companies. Their opinions have also been of great help in improving the system, and their assessment very satisfactory.

USALSIM is a project with a clear goal of improving and continuing the use of this tool. New objectives and expansions have been defined for the next few months, at both the technical and content level, as well as the inclusion of other possible branches of knowledge, the integration of more institutions and business entities to provide work placements, etc., giving this platform a strong projection toward the future.

References

1. Schroeder, R.: Possible worlds: the social dynamic of virtual reality technologies. Westview Press, Boulder (1996)
2. Schroeder, R.: Defining Virtual Wolrds and Virtual Environments. Journal of Virtual Worlds Research 1(1) (January 2008)
3. Kolb, D.A., Boyatzis, R.E., Mainernelis, C.: Experimental learning theory: Previous research and New Directions. In: Sternberg, R.J., Zhang, L.F. (eds.) Perspectives on Thinking, Learning and Cognitive Styles, pp. 227–248. Lawrence Erlbaum, Mahwah (2002)
4. Barab, S., Thomas, M., Dodge, T., Carteaux, R., Tuzun, H.: Making Learning Fun: Quest Atlantis, A Game Without Guns. Educational Technology Research and Development 1, 86–107
5. Dalgarno, B., Bishop, A.G., Adlong, W., Danny, J., Bedgood, R.: Effectiveness of a Virtual Laboratory as a preparatory resource for Distance Education chemistry students. Comput. Educ. 53, 853–865 (2009)
6. OpenSimulator, http://www.opensimulator.org
7. Tanenbaum, A.S., Van Steen, M.: Sistemas Distribuidos, principios y paradigmas, 2nd edn. Pearson Educación, México (2008)
8. Moodle, http://www.moodle.org
9. Sloodle, http://www.sloodle.org
10. Konstantinidis, A., Tsiatsos, T., Demetriadis, S., Pomportsis, A.: Collaborative Learning in OpenSim by Utilizing Sloodle. In: 2010 Sixth Advanced International Conference on Telecommunications (AICT), pp. 90–95 (2010), doi:10.1109/AICT.2010.75
11. Kemp, J.W., Livingstone, D., Bloomfield, P.R.: SLOODLE: Connecting VLE tools with emergent teaching practice in Second Life. British Journal of Educational Technology 40, 551–555 (2009), doi:10.1111/j.1467-8535.2009.00938.x

How to Design Games for Deaf Children: Evidence-Based Guidelines

Alessandra Melonio and Rosella Gennari

Abstract. The goal of this paper is to present the first evidence-based guidelines for the design of electronic games for deaf children. According to the most recent deaf literature, playing with such games shows positive effects on deaf children's visual abilities and working memory abilities. Our review of deaf literature, briefly sketched in the paper, considers such abilities as well as other relevant findings concerning the needs of deaf children most relevant for the design of electronic games for them. The paper also outlines the latest findings of the TERENCE project, which builds electronic smart games for deaf children. All such findings are then use to compile the guidelines, which are presented in the third and final part of this paper.

Keywords: evidence-based design, user centred design, deaf studies, games, children with special needs, usability and accessibility.

1 Introduction

In recent years, more and more attention is being paid to the design of *electronic tools* (e-tools) for children, and there is a fair amount of work in which designers have started developing design principles for e-tools for children (e.g. in [21,34]). To the best of our knowledge, however, there is no single collection of principles for the design of e-tools for deaf children. We found a list of suggestions for evaluating e-tools for deaf people [28] and guidelines for captioning for web sites for them [29]. On the other hand, the benefits of e-tools for the deaf population are purported by deaf research studies. In particular, recent deaf studies show how playing video games can have positive effects in terms of visual abilities and working memory of deaf individuals [14;16]. Therefore we set ourselves on such a tack: the main goal of this paper is to present the first evidence-based guidelines for the design of electronic games that are accessible and usable for deaf children.

Alessandra Melonio · Rosella Gennari

KRDB, Free University of Bozen-Bolzano, P.zza Domenicani 3, 39100 Bolzano

e-mail: alessandra.melonio@stud-inf.unibz.it,
        rosella.gennari@unibz.it

P. Vittorini et al. (Eds.): *2nd International Workshop on Evidence-Based TEL*, AISC 218, pp. 83–92.

DOI: 10.1007/978-3-319-00554-6_11   © Springer International Publishing Switzerland 2013

We start with a compact overview of most relevant deaf studies for deaf individuals, focusing on deaf children. The overview highlights what we know and we do not know from the literature about the characteristics of deaf people and relevant for designing games for deaf children. With the goal of learning more about such an issue, we conducted experiments with deaf children, their teachers and experts of deafness for the TERENCE European project [23], which is developing video games for improving the reading comprehension of children, like deaf children. In particular, the TERENCE consortium run field studies with children as subjects and their referent adults as informants. We designed the tasks of the field study with children as paper-and-pencil games, collected the results of the games and also observed children while playing with them. The state-of-the-art analysis and the results of the field studies run for TERENCE, allow us to compile a set of guidelines for the design of games for deaf children, which is the focus of the third and final part of the paper.

2   Research Findings

In this section, we analyse the most relevant needs of deaf children for playing video-games and, mainly, concerning reading, attention and memory. The needs emerge from an analysis of the deaf literature and recent findings of the TERENCE project [23]. TERENCE is developing an adaptive learning system for improving the reading comprehension of primary-school children, hearing and deaf, by means of stories and smart games for reasoning about stories. In order to understand the needs of children for reading and playing with the TERENCE system, the TERENCE consortium run studies following the *user-centred design* (UCD) [18]: the consortium conducted expert-based studies, with experts of the domain or UCD as participants, and user-based studies, in which the participants were children, hearing and deaf, and their referent adults, like class teachers, support teachers and parents. The studies were done firstly for (1) the context of use analysis and secondly for (2) the evaluation of prototypes of the system. The first studies were done for analysing the impact on the design of the system (a) of the characteristics of the users, (b) of the tasks they can perform with the system like playing computer games, and (c) of the environment. The data collection involved 592 7–11 olds across UK and Italy, 70 out of which are deaf, and about 30 referent adults, that are parents of children, class teachers or support teachers. Data collection activities with children were in the form paper-based games, and data collection with adults was done via contextual inquiries, questionnaires or diaries. Direct observations complemented all data collection activities in situ. See [24]. The results were picked up for designing the TERENCE system, in particular, its smart games and the related interface for playing with them. The resulting high-fidelity prototype, realised in Flash, was then evaluated in the second studies, that is, usability testing sessions of c.a 1 hour each. Tasks with the prototype were analysed

Table 1 Reading

	CONSTRAINT	BIB	GL #
Deaf studies	- Word recall by deaf seems poor for long words, as well as for abstract, ambiguous or unfamiliar words without contextual clues. - Deaf children's vocabulary skills are better when words have only a single meaning or when they are presented in context. Unfamiliar words, or words which have not been specifically introduced to the student, cannot be lip-read. - Reading involves using of the centre of visual field to fixate the word for hearing children. Therefore the fact that deaf children pay more attention to items in the periphery could partially cause confusion in the identification of letters and words.	[1;2;7;13;30;31;35]	3.1.1 3.1.2 3.1.3 3.2.2 3.2.3
	- Deaf individuals seem to have problems with complex sentences, in particular, with cohesive devices and referential expressions. - Deaf students tend to remember disconnected portions of the text rather than the whole picture, especially when the material is unfamiliar.	[24;25;31;35]	3.2.1 3.2.3
	- Deaf readers, like good hearing readers, use metacognitive strategies to monitor and maintain comprehension, but are less accurate in their meta-comprehension. - Deaf readers seem to benefit from a "windowed reading" whereby only limited amounts of text are made available at any one time	[19;17]	3.2.2 3.2.3
TERENCE	- They prefer reading short texts and books with pictures. - When deaf children are reading books, the teacher often has to recall the attention of the children and indicate the point where they were reading.. - Deaf children are likely to have problems with: tapping global coherence as well as local cohesion; complex periods and, in particular, co-references; decoding; phonology. - Instructions are not always read; deaf children read them only if they appear before the start of the activity.		3.1.2 3.2.2 3.2.3 3.2.4

Table 2 Visual attention

	CONSTRAINT	BIB	GL #
Deaf studies	- Deafness leads to changes not in all aspects of vision but specifically in visual attention and alteration of attention abilities. - Deaf individuals seem better in certain aspects of visual perception and specifically at allocating visual attention to the periphery of the visual field. - Deaf signers seem to be more distracted by peripheral events and hearing individuals are more distracted by central events.	[1;2;3;5;7;17;36;37;]	3.2.2 3.2.3 3.3.1
	Young children have more difficulties for serial recall and take more time for recovering attention.	[14]	3.2.2
	Deaf individuals are better than hearing individuals in *orienting visual attention* from one location to another, and are more affected by the presence of distractors, that is, they are less good in *selective attention*, whereas no difference was found in *divided attention* that is the ability of processing multiple stimuli in the visual field.	[6;7]	3.2.3 3.3.1 3.3.2 3.3.3
	In deaf individuals the ability to discriminate very small differences in direction of motion is altered and more deaf subjects discriminated gross differences in direction as leftward vs rightward.	[12]	3.3.3

Table 3 Focus and Social Interaction

	CONSTRAINT	BIB	GL #
Deaf studies	The majority of deaf children have problems in focussing their attention.	[4]	3.4.1 3.4.3
	Few mothers declare that they have problems in eliciting and maintaining eye gaze and joint attention with their deaf children.	[38]	3.4.2 3.4.3
	Deaf children tend to be more impulsive and lack of inhibition and suffer for increased distractibility.	[8;39;40]	3.4.1
	Calibrated use of vibration feedback or motion for deaf children may be use to get their attention focussed.	[20]	3.4.1
	Children's performance in WM was directly related to the number of games that they played. Several studies show that children who play action video games showed enhanced performance on all aspects of attention.	[15;22; 33]	3.5.6
TERENCE	- The deaf child is distracted more easily and should always be called his/her attention with signs. When deaf children are reading books, the teacher often has to recall the attention of the children and point to where they were reading. - Deaf children tend to have diminished attention time (after a bit they are bored). - If the teacher uses pictures or games the deaf child is more stimulated to perform reading tasks. - They are more alert of being treated differently and suffer from it. The older they become, the more frustration-prone they grow.		3.3.2 3.4 3.5.3 3.5.4 3.5.5
	Deaf children devote less time to cooperative activities and significantly more time to solitary activities.	[41]	3.4.3 3.5.1

Table 4 Memory

	CONSTRAINT	BIB	GL #
Deaf studies	Reading ability is closely linked to overall short-tem memory performance. This seems lower for deaf individuals, so is long-term memory.	[9;11]	3.1 3.2
	- Deaf children surpass hearing children in short-term memory tasks for complex figures, except when the task involved serial recall. - Deaf people are accredited to rely more heavily on visuo-spatial short-term memory codes. For instance, deaf subjects have deficits in recall for linguistic stimuli, printed words and pictures but not in recall of non-linguistic stimuli such as unfamiliar faces and spatial arrays of lights.	[9;10; 12;14]	3.3.1 3.3.3 3.4.3 3.5.3 3.5.4
	Deaf individuals may be a disadvantage on linguistic tasks that involve serial recall but they seem to be better in tasks that involve temporal order.	[12;32]	3.3.1 3.5.4
	Children's performance in memory tasks seems directly related to the number of games that they play: the more games they play, the better their performance in memory tasks when retested.	[6;9]	3.5
TERENCE	- Deaf children show to better recall images than texts alone. - Since their first impact is with the physical aspect of a person, they tend to remember this better. Often, they refer to person by signing physical characteristics of the person, e.g., the curly girl.		3.2.2 3.4.2

Table 5 Types of Games

	CONSTRAINT	BIB	GL #
Deaf	In a variety of problem solving and other academic tasks, deaf students have been found more likely than hearing age-mates to focus on individual item information rather than relations among items.	[16;43]	3.5.2 3.5.5
TERENCE	Deaf children generally prefer: - human-like avatars to others; - non-photorealistic consoles; when the age increases the use of non-photorealistic consoles decreases and the use of photorealistic consoles increases - portable devices (e.g. Nintendo DS or tablets); - games of movement (they often refer to the usage of balance board); - playing by themselves, alone, and in the same place.		3.5

in terms of their success and observations allowed us to detect unique usability issues. The results of all the studies of TERENCE for deaf children are in-line with those found in the literature. Table 1, 2, 3, 4 and 5 recap the findings. Each table is related to specific characteristics (e.g. reading, attention, etc). A table is structured into two main parts: the white part is related to deaf studies; the blue part is for TERENCE findings alone.

3 Guidelines for the Design of Usable Games for Deaf Individuals

There are many studies about design principles for technology artefacts for hearing children, but not for deaf children. In this section we state the first guidelines for designing computer games that are accessible for deaf children, and that arise from the research in deafness reported above. We clustered the guidelines into the following 5 main areas that are explained below.

3.1 Words on the Screen

According to the literature review, unfamiliar or ambiguous words, without contextual clues, are problematic for deaf children and words that have not been specifically introduced to the student cannot be lip-read. Moreover, if deaf readers have an alteration in their visual selective attention, they could have problems in identifying the letters of a word and in creating representations that preserve both the correct letters and their correct spatial arrangements. Therefore texts should use *familiar and unambiguous words, paying attention to neighbouring words that influence where the reader will fixate their attention [GL 3.1.1].* If unfamiliar, ambiguous or abstract words are used *then their meaning should be easy to be inferred from the surrounding context [GL 3.1.2].* As explained in the literature review, word length matters, thus *words should not be too long [GL 3.1.3].*

3.2   Other Characteristics and Position of Text

Reading problems, issues with attention and memory suggest several guidelines for how texts should be positioned in screens for playing games. This is particularly true for instructions for playing games; accessible instructions accelerate the time to perform the playing task. Firstly, according to what in Sect. 2.1, any *explanatory text*, as in *instructions, should use short and simple sentences, without complex referential expressions [GL 3.2.1]*. For instance, deaf children will have problems to resolve who "her friend is" in the sentence "One day Ben and Sophie visited the biggest swimming pool in town with their Mum and her friend from work". Moreover, given the visual attention orienting and selective skills of deaf learners, the game should *use visual clues or animations for directing the attention of the child on relevant textual information [GL 3.2.2]*. Moreover, *relevant textual information* like instructions *should occur in a separate dedicated part of the screen, with a small amount of information* because deaf children may have problems with longer fixation and slower reading times and, in general, they perform better if limited amounts of text are made available one at a time *[GL 3.2.3]*. *Instructions* should better be placed *before the start of the game*, as suggested by our usability studies *[GL 3.2.4]*.

3.3   Characteristics and Positions of Other Objects on the Screen

According to the literature review of this paper on attention, young children have more difficulties for serial recall and take more time for recovering attention. This means that *younger learners may need fewer choices than older children in games*. More in general, *using the same items in the same position and order in the interface should aid the recall of deaf children [GL 3.3.1]*. On the screen, there should not be distracting stimuli for the peripheral visual field of view because deaf individuals are more distracted by peripheral events. *On the edge of the screen, the interface should have objects and motion stimuli that do not distract the children from their main task [GL 3.3.2]*. Deaf individuals are better than hearing in their ability to orient spatial attention especially at reorienting it from one location to another. More deaf subjects discriminate gross differences in direction as leftward versus rightward. This means that the interface of the *game should use the motion of objects only in relation to the main task for the children [GL 3.3.3]*.

3.4   Interaction and Feedback

In general, children are impatient and need immediate feedback: they expect to see the results of their actions immediately; if nothing happens after their input, children may repeat their action until something does occur. Deaf children are problems to focus attention for too a long time in a reading task or demanding playing activities. In general, a child should not be left idle in front of the screen for too long a time without any stimulus or feedback. The game for deaf children, thus, could have *vibration or motion feedback for directing the attention [GL 3.4.1]* of the learner towards specific targets, e.g., the correct or wrong resolution of a game. However, one must be careful where to place the animation on the screen because it might adversely affect their focus attention. Since deaf children are more impulsive, *the type of feedback must be calibrated on the target deaf children [GL 3.4.2]* so as not to be frustrating or irritating. While hearing children can listen and answer simultaneously within the game, *deaf children must interact with one task at a time, e.g., the game should propose a reading task and a resolution task in separate moments [GL 3.2.3]*.

3.5 Game Genres and Avatars

According to the literature review, possibly due to difficulties in communicating and socially interacting with nearby peers, *deaf children prefer single-player games [GL 3.5.1]*. In our usability study with deaf and hearing children, we observed that all our children's preferences were for *playing with consoles* (about 27% of preferences). All children prefer doing specific activities always in the same place. The majority of deaf children prefer playing with consoles alone and prefer *games with movement [GL 3.5.2]* (e.g. balance board of WII or kinect of XBOX). Therefore they need sufficient space to move freely while playing. Deaf children often fail to respond with gestures or signs when their eyes are attracted by the objects in motion, due to their difficulty with divide attention. While hearing children can listen and answer simultaneously within the game, deaf children must interact with one task at time. Moreover deaf children suffer from increased distractibility and have different attention needs according to the literature. Therefore the *duration of the game should not be too long and composed of a single task at a time [GL 3.5.3]*. However, since deaf children are easily irritated, the *timing of games should be calibrated on the target deaf players [GL 3.5.4]*. Deaf children perceive immediately when they are treated differently, the older they grow and the less impatient they become. Thus, *the game should pay special attention to the age of the child, e.g., the genre of texts and pictures should be age-appropriate [GL 3.5.5]*. Several studies show that children who played action video games showed enhanced performance on all aspect of attention. Moreover, from our own usability experiments with deaf and hearing children, it turns out that playing with video games takes a large part of the deaf children's day, and is preferred over other daily activities (e.g. TV, reading). So *the training with games like action games may be used to enhance deaf children's skills, in particular, for improving their performance problem solving strategies [GL 3.5.6]*, possibly enhancing their working memory. According to our studies, *human-like avatars guiding through games [GL 3.5.7]* were the most appreciated.

Acknowledgments. The authors' work was supported by TERENCE project, funded by the EC through the FP7 for RTD, Strategic Objective ICT-2009.4.2, ICT, TEL.

References

1. Bavelier, D., Dye, M., Hauser, P.: Do deaf individuals see better? Trends in Cognitive Sciences 10(11), 512–518 (2006)
2. Proksch, J., Bavelier, D.: Changes in the spatial distribution of visual attention after early deafness. Journal of Cognitive Neuroscience 14(5), 687–701 (2002)
3. Smith, L.B., Quittner, A.L., Osberger, M.J., Miyamoto, R.: Audition and visual attention: The developmental trajectory in deaf and hearing populations. Developmental Psychology 34(5), 840–850 (1998)
4. Dye, M.W.G., Hauser, P.C., Bavelier, D.: Visual Attention in Deaf Children and Adults. Implications for Learning Environments. In: Marschark, M., Hauser, P.C. (eds.) Deaf Cognition: Foundations and Outcomes, pp. 250–263. Oxford University, New York (2008)
5. Dye, M.W.G., Bavelier, D.: Differential development of visual attention skills in schoolage children. Vision Research 50(4), 452–459 (2010)
6. Bosworth, R.G., Dobkins, K.R.: The effects of spatial attention on motion processing in deaf signers, hearing signers and hearing non signers. Brain Cognition 49(1), 152–169 (2002)
7. Dye, M.W.G., Hauser, P.C., Bavelier, D.: Is visual selective attention in deaf individuals enhanced or deficient? The case for the Useful Field of View. PLoS ONE 4(5), e5640 (2009), doi:10.1371/journal.pone.0005640
8. Quittner, A.L., Leibach, P., Marciel, K.: The impact of cochlear implants on young deaf children: New methods to assess cognitive and behavioral development. Archives of Otolaryngology and Head and Neck Surgery 130(5), 547–554 (2004)
9. Marschark, M., Mayer, T.S.: Mental Representation and Memory in Deaf Adults and Children. In: Marschark, M., Clark, D. (eds.) Psychological Perspectives on Deafness, vol. 2, pp. 53–77. Lawrence Erlbaum and Associates, Mahwah (1998)
10. Todmann, J., Seedhouse, E.: Visual action code-processing by deaf and hearing children. Language and Cognitive Processes 9, 129–141 (1994)
11. Macsweeney, M., Campbell, R., Donlan, C.: Varieties of short-term memory coding in deaf teenagers. J. Deaf Stud Deaf. Educ. 1(4), 249–262 (1996)
12. Todmann, J., Cowdy, N.: Processing of visual attention codes by deaf and hearing children: Coding orientation or M-capacity? Intelligence 17, 237–250 (1993)
13. Campbell, R., Wright, H.: Deafness and immediate memory for pictures: Dissociations between 'inner speech' and the 'inner ear'. Journal of Experimental Child Psychology (1990)
14. Grigonis, A., Narkevičienė, V.: Deaf Children's Visual Recall and Its Development in School Age. Vytauro Didžiojo universitetas K. Donelaičio g. 52, Kaunas (2010)
15. Nunes, T., Evans, D., Barros, R., Burman, D.: Can deaf children's working memory span be increased? Department of Education. University of Oxford
16. Marschark, M., Everhart, V.S.: Problem-solving by deaf and hearing students: twenty questions. Deafness Educ. Int. 1, 65–82 (1999)
17. Dye, et al.: Visual skills and cross-modal plasticity in deaf readers: possible implications for acquiring meaning from print. Ann. N Y Acad. Sci. 1145, 71–82 (2008)

18. Gulliksen, J., Göransson, B., Boivie, I., Blomkvist, S., Persson, J., Cajander, Å.: Key principles for user-centred systems design. Behaviour & Information Technology 22, 6 (2003)
19. Gibbs: Individual differences in cognitive skills related to reading ability in the deaf. American Annals of the Deaf 134(3), 214–8 (1989)
20. Hertzog: Categorization of Vibration Feedback at Different Levels: A Study with Deaf and Hard-of-Hearing Consumers. 011 RIT Summer Undergr. Research Symposium (2011)
21. Design Principles for Children's Technology. Sonia Chiasson and Carl Gutwin. Department of Computer Science, University of Saskatchewan. HCI-TR-2005-02 (2005)
22. Buckley, D., Codina, C., Bhardwaj, P., Pascalis, O.: Action video game players and deaf observers have larger Goldmann visual fields. Vision Research 50, 548–556 (2010)
23. TERENCE Project. Website (2011), http://www.terenceproject.eu
24. Di Mascio, T., Gennari, R., Melonio, A., Vittorini, P.: The user classes building process in a TEL project. In: Vittorini, P., Gennari, R., Marenzi, I., de la Prieta, F., Rodríguez, J.M.C. (eds.) International Workshop on Evidence-Based TEL. AISC, vol. 152, pp. 107–114. Springer, Heidelberg (2012)
25. Trezek, B.J., Paul, P.V., Wang, Y.: Reading and deafness: Theory, research, and practice. Delmar, Cengage Learning, Clifton Park (2010)
26. Traxler, C.B.: The Stanford Achievement Test, 9th Edition: National norming andperformance standards for deaf and hard-of-hearing students. Journal of Deaf Studies and Deaf Education 5(4), 337–348 (2000)
27. Marschark, M., Convertino, C.M., Macias, G., Monikowski, C.M., Sapere, P., Seewagen, R.: Understanding Communication among Deaf Students Who Sign and Speak: A Trivial Pursuit? American Annals of the Deaf 152, 415–424 (2007a)
28. Mich, O.: Evaluation of software tools with deaf children. In: Proceedings of the 11th International ACM SIGACCESS Conference on Computers and Accessibility, pp. 235–236 (2009)
29. Accessible web sites, http://www.samizdat.com/pac2.html
30. Marschark, M.: Education and development of deaf children: Or is it development and education? In: Spencer, P., Erting, C., Marschark, M. (eds.) Development in Context: The Deaf Child in the Family and at School, pp. 275–292. LEA, Mahwah (2000)
31. Marschark, M.: Language development in children who are deaf: A research synthesis. National Association of State Directors of Special Education, Alexandria (in press)
32. Marschark, M.: Psychological Development of Deaf hildren. Oxford University Press, Oxford
33. H.-Wallander, B., Green, C.S., Bavelier, D.: Stretching the limits of visual attention: the case of action video games
34. Grammenos, D., Paramythis, A., Stephanidis, C.: Designing the User Interface of an Interactive Software Environment for Children. Institute of Computer Science, Foundation for Research & Technology – Hellas Science and Technology Park of Crete, Heraklion, Crete

35. Banks, J., Gray, C., Fyfe, R.: The written recall of printed stories by severely deaf children. British Journal of Educational Psychology 60, 192–206 (1990)

36. Bosworth, R.G., Dobkins, K.R.: The effects of spatial selective attention on motion processing in deaf and hearing subjects. Brain & Cognition 49(1), 170–181 (2001)

37. Fine, I., Finney, E.M., Boynton, G.M., Dobkins, K.R.: Comparing the effects of auditory deprivation and sign language within the auditory and visual cortex. Journal of Cognitive Neuroscience 17(10), 1621–1637 (2005)

38. Meadow, K.P.: Deafness and child development. Univ. of CA Press, Berkeley (1980)

39. Quittner, A.L., Smith, L.B., Osberger, M.J., Mitchell, T.V., Katz, D.B.: The impact of audition on the development of visual attention. Psychological Science (1994)

40. Reivich, R.S., Rothrock, I.A.: Behavior problems of deaf children and adolescents: A factor-analytic study. Journal of Speech and Hearing Research 15, 84–92 (1972)

41. Higginbotham, D.J., Baker, B.M.: Social participation and cognitive play differences in hearing impaired and normally hearing preschoolers. The Volta Review 83 (1981)

42. Cain, K.: Making sense of text: skills that support text comprehension and its development. Perspectives on Language and Literacy 35, 11–14 (2009)

43. Marschark, M., Convertino, C., LaRock, D.: Optimizing academic performance of deaf students: Access, opportunities, and outcomes. In: Moores, D.F., Martin, D.S. (eds.) Deaf Learners: New Developments in Curriculum and Instruction, pp. 179–200. Gallaudet University, Washington, D.C (2006)

Simulation Still Rules: Modern Computer Systems Architecture Visualized and Laid Bare

Besim Mustafa

Abstract. Software simulation is an invaluable and indispensable educational tool in teaching and learning of complex systems' behavior specially when accompanied with effective visualization and animation methods. Examples of simple but effective visualization and animation methods are presented with reference to a unique set of tightly-integrated simulators designed to engage and capture the imagination of students of modern computer architecture. Several years of practical experience and development are largely based on user feedback and evaluation work that are indicative of the positive impact of the simulations on enhancing learning experiences of the students from basic to advanced levels of study.

1 Introduction

Software simulation is a well know method of studying the behavior of systems using computer software and has been in use for a long time in many diverse areas from research through product design to entertainment and accident investigations thus serving as an invaluable and indispensable tool; it still continues to be so.

In this paper we describe the simulation methods we used to 'bring to life' different architectural features of modern computer systems by way of simple but effective visualizations and animations for educational purposes. Today's students with learning styles that favor visualizations and animations find simulations engaging, helpful and supportive in their studies. The visualizations we present as examples are implemented by our unique integrated system simulator designed to support our teaching. The novelty in our simulations lies primarily in the way the key concepts are covered and the interdependencies between them are explored. We took the decision to design our own simulator when we failed to find a simulator that is able to help visualize the 'big picture' and demonstrate the way different parts of a system 'hang' together; both attributes useful in enabling deeper understanding. Thus the emphasis of the paper is on the visualization techniques rather than the usability issues that often concern themselves with user interactions.

Besim Mustafa

Department of Computing, Faculty of Arts and Sciences, Edge Hill University, Ormskirk, UK

e-mail: mustafab@edgehill.ac.uk

P. Vittorini et al. (Eds.): *2nd International Workshop on Evidence-Based TEL*, AISC 218, pp. 93–100.

DOI: 10.1007/978-3-319-00554-6_12   © Springer International Publishing Switzerland 2013

2   The System Simulator

We very briefly describe the key features of our integrated system simulator as background information; it is comprised of a CPU simulator, an OS simulator and a high-level 'teaching language' compiler [1] all integrated in one single package. The tutorial notes guide our students to becoming familiar with and proficient in the simulators in progressive stages throughout their three-year degree course.

The CPU simulator simulates RISC type architecture; it has a small set of CPU instructions, a large register file and incorporates a five-stage pipeline simulator as well as Harvard style separate data and instruction cache simulators.

The OS simulator supports two main aspects of a computer system's resource management: process management and memory management. The process scheduler supports several scheduling mechanisms. Threads, mutual exclusion, process synchronization and deadlock concepts are explored via teaching language constructs; the virtual memory simulations explore the principles of address translation, paging, placement and replacement concepts.

The 'teaching' compiler demonstrates the generation of low-level CPU instructions and explores key instruction optimization methods.

3   The Visualizations

In this section we look at examples of visualization techniques used in our simulations. As over 90% of our students are visual learners we feel justified in our efforts in this area. While the simulations support and complement the theory they also need to be engaging and involving. To this end we therefore suitably adopted the principles identified in the 'Taxonomy of Engagement' [2] by the working group on the impact of algorithm visualizations while paying careful consideration to 'Blooms Taxonomy' of educational objectives [3]. Our efforts in this are further explored in [4] where we present our evaluation methods and the results.

3.1   CPU Simulator Visualizations

There are three areas of CPU architecture that particularly lend themselves to visualization and animation: instruction execution, caching and pipelining technologies. The CPU simulator demonstrates the main functions of a typical CPU by employing several visualization methods. It does this by presenting views of instructions in memory, data in stack and in CPU registers as shown in Figure 1. The execution of CPU instructions are visually enhanced by highlighting instructions as they are executed and by animating the instances of updating data in registers and on the program stack as and when these areas are accessed.

Advanced CPU simulator features include a 5-stage pipeline simulator, data and instruction cache simulators. The individual stages of the pipeline are represented as blocks of different colors. The pipeline can be switched on or off in order to visually demonstrate parallel execution as opposed to sequential execution of instructions. Figure 2 shows the stages of instructions as they go through the pipeline in serial sequence whereas figure 3 depicts parallel execution visually highlighting the difference between the two as the instructions are executed.

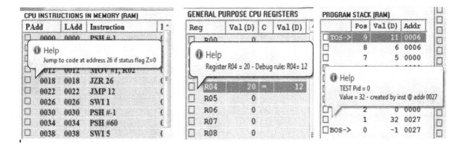

Fig. 1 CPU simulator visualizations: instructions (left), registers (middle) and stack (right)

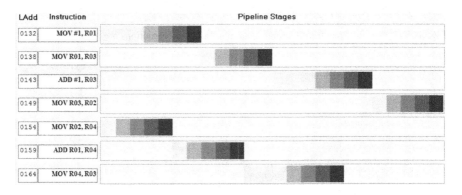

Fig. 2 CPU pipeline executing instructions in series, i.e. no instruction pipelining

The pipeline hazards such as data dependencies and control hazards are also represented as color-coded blocks. Figure 3 (top left) shows a section of the pipeline display where the red blocks (or the 1st dark block in 2nd and 4th to 6th lines in gray-scale images) represent data hazards due to data dependencies between the instructions. Figure 3 (top right) shows the same sequence of instructions but this time with operand forwarding optimization enabled. Here the data hazards are effectively mitigated by operand forwarding as indicated by the blocks labeled FW. The performance improvement is clear in the displayed statistics in Figure 3 (bottom images); same number of instructions is executed in less clock cycles (i.e. 18 instead of 20). Another advanced feature included is the simulation of data and instruction caches. As seen in Figure 4, the structure and the contents of the caches as well as the graphical and animated views of the miss rates are displayed for different cache architectures and sizes allowing students to study the differences.

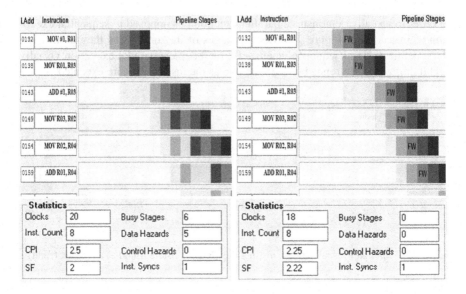

Fig. 3 CPU pipeline executing instructions in parallel, i.e. instructions are pipelined

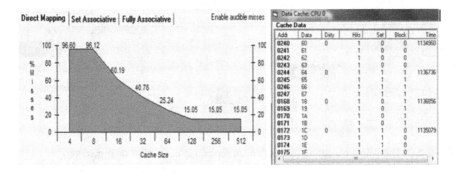

Fig. 4 Plot of CPU cache misses against different cache sizes (left); cache data display (right)

3.2   OS Simulator Visualizations

The OS simulator provides additional visualization challenges requiring visual representation of system management of resources such as processes and memory involving process states, state transitions, placement, replacement, deadlock detection and resolution as well as load balancing with CPU/memory utilization.

Figure 5 shows two versions of graphical representation of process states. The image on the right is an animated version of state transitions as the processes are dispatched, executed or blocked. In the meantime, Process Control Blocks (PCBs) facilitate context switching between processes an example of which is shown in Figure 6 (right). Memory management is needed to share and protect finite memory amongst multiple processes. The OS simulator supports paged memory management simulation and space permitting the pages are placed in memory otherwise swapped out. Figure 7 (left) shows the page table for process 11 where the physical memory addresses, the frames and the page fault counts are shown. Figure 7 (right) depicts the pages of process 11 in main memory where each process's pages are assigned colors and virtual activity is animated as the process pages are swapped in and out with visually enhancing color code changes. Modern operating systems provide support for multithreading without which server technology would not be possible. The OS simulator supports simulation of multithreading through high-level language constructs. Figure 8 (left) shows the tree structure of threads clearly illustrating all important parent/child relationships. The image on the right shows the corresponding high level source supporting threads.

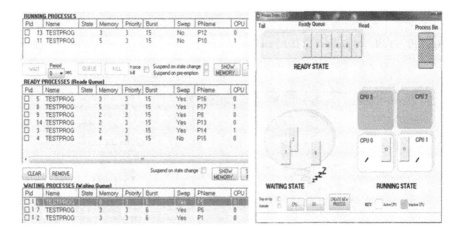

Fig. 5 Two different views of the same process states; the image on the right is animated

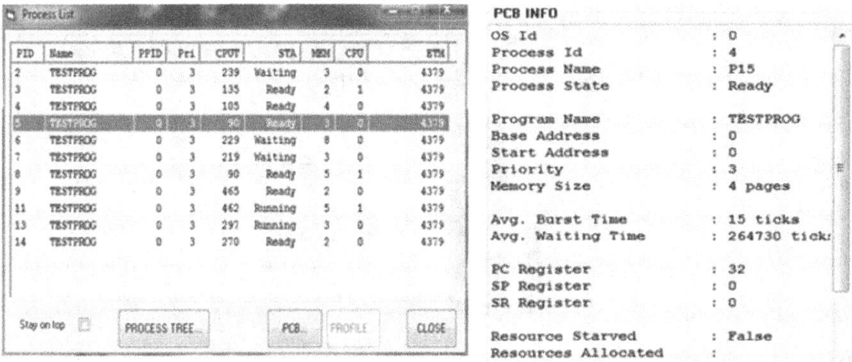

Fig. 6 Process listing with periodic updates of states (left) and example PCB display (right)

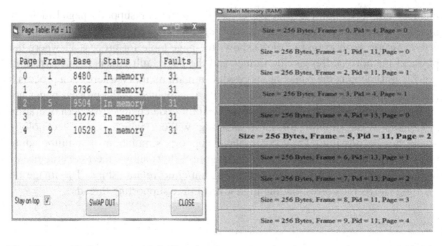

Fig. 7 Page table for process 11 (left) and main memory showing pages of process 11 (right)

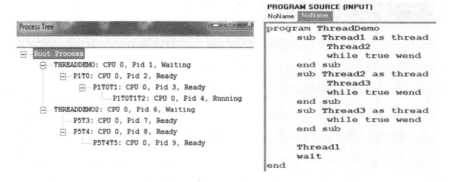

Fig. 8 Display of process trees (left) and the source corresponding to the top tree (right)

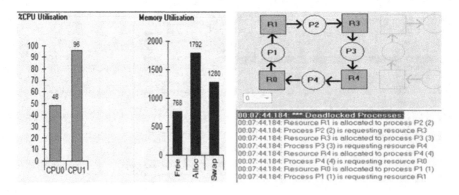

Fig. 9 CPU and memory resource utilization (left) and deadlock detection (right)

Figure 9 (left) is a snapshot of resource utilization display with two CPUs sharing the processes and can be used to demonstrate and explore CPU load balancing. A consequence of managing multiple processes and finite resources is the possibility of deadlocking. The OS simulator incorporates a deadlock simulator that facilitates the studying of deadlock prevention, detection and resolution techniques. Figure 9 (right) is a graphical representation of deadlocked processes P1 to P4.

3.3 'Teaching Compiler' Visualizations

The teaching compiler compiles high-level language statements into assembly instructions and the equivalent byte code while at the same time displaying different stages of compilation process as shown in Figure 10. It is also possible to animate the tracing of source lines corresponding to the CPU instructions as they are run.

The source editor is at top left; below that is the display of different stages of the compilation process. The top right image shows the CPU instructions in assembly language format and below that is the byte code equivalent. The students can highlight a source statement whereupon the corresponding code is highlighted and vice versa. Selected compilation progress displays are associated with corresponding source statements facilitating understanding as can be seen in Figure 10.

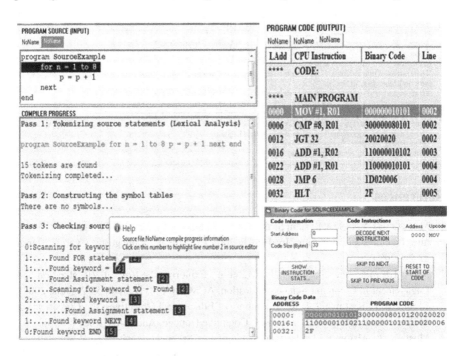

Fig. 10 Example source, corresponding compiler progress display and code generated

4   What the Users Say and Think

Contents of Table 1 lend some justification to the effectiveness of the simulations and the visualizations we adopted. Further qualitative and quantitative evaluations we conducted [4] also add weight to the positive impact of our methods. We continue to improve our visualizations and made our simulators public via a dedicated educational site [5] that has been receiving much International interest.

Table 1 A small sample of student reflections and comments from instructors on the simulations

What students say:
"Being able to see scheduling actually working on the simulator made it a lot easier to understand"
"The lecture confused me slightly about threads however once I had used the simulator I understood it more"
"Today's session was interesting, it was a change actually opening a simulator and physically loading processes into the operating system"
"The simulator made the theory much more understandable"
"The knowledge gained from the lecture and the help from the simulator made this session comfortable and easy to complete"
"Today's session has helped me to develop a greater understanding of process scheduling and this was due to the simulator"
"Today's session was particularly enjoyable as it allowed for group work and communication as well as seeing how the process scheduling works within an operating system by using the simulator"
"I feel that I have learnt the idea of threads much easier by actually seeing and understanding the ways in which they work as well as being able to discuss these findings with group members"
"The simulator let me see first-hand what happens exactly"
"I really enjoyed using the CPU Simulator and learning how it worked"

What instructors say:
"I found your YASS Simulator very friendly and I'm planning to use it on this semester on my Computer Science class." (A college teacher from Brazil)
"I played with it a little yesterday and see a lot of opportunities." (A college instructor from North Carolina)
"I came across your CPU-OS simulator via the Jorum website, and I would like to consider using it for a computer systems module I am associated with." (A university lecturer from Scotland)
"I have just found your simulator.... I am going to teach a computer architecture course this semester and looking for a simulator." (A university lecturer from Sri Lanka)
"I sent you this e-mail just to say how I appreciate your work on CPU-OS simulator." (A research student from Turkey)
"I'm very pleased to have found your simulator.... It's very useful as far I can see until now." (A university lecturer from Chile)
"I am teaching an OS course.... It is quite interesting your simulator." (A college instructor from Canada)
"I have come across your cpu simulator which I would like to use to teach my students some basic machine operations and support the lecturing notes on cpu architecture." (A college instructor from England)
"I am a teaching assistant of the computer organization.... it is a very good simulator especially for students to understand what pipeline is." (A university lecturer from Taiwan)

References

1. Mustafa, B.: YASS: A System Simulator for Operating System and Computer Architecture Teaching and Learning. In: FISER 2009 Conference, Famagusta, Cyprus, March 22-24 (2009)
2. Naps, T.L., Fleischer, R., McNally, M., et al.: Exploring the Role of Visualization and Engagement in Computer Science Education. ACM SIGCSE Bulletin 35(2) (June 2003)
3. Bloom, B.S., Krathwohl, D.R.: Taxonomy of Educational Objectives; the Classification of Educational Goals. In: Handbook I: Cognitive Domain. Addison-Wesley (1956)
4. Mustafa, B., Alston, P.: Understanding Computer Architecture with Visual Simulations: What Educational Value? In: International Workshop on evidenced-based Technology Enhanced Learning, Salamanca, Spain, March 28-30 (2012)
5. Teach-Sim educational simulators, http://www.teach-sim.com (accessed January 10, 2013)

How to Get a Good Job and Improve Your Career – With a PLE

Luigi Romano

Abstract. Personal Learning Environments (PLEs) are gaining increasing interest because they provide a flexible way to better match students' needs. The main in using widely a PLE is the danger of the owner being unable to discriminate sound contents from groundless or wrong ones. The paper describes an experience where students were taught how to build a PLE in order to strengthen their understanding of Mathematics. The paper focuses on two aspects: fostering and strengthening the creation of critical sense in teenaged students when building their PLE, and overcoming personal difficulties and misunderstandings about Mathematics with a PLE. Gaining such skills will be extremely useful for improving students' personal and working life, in a Lifelong Learning perspective.

Keywords: Personal Learning Environment, Critical Sense, Lifelong Learning, web based Learning, Communities of Practice, OECD-PISA, Mathematics.

1 Introduction

Massive adoption of computers in Educational activities, together with the pervasive use of Internet, make it possible nowadays to access huge amounts of learning materials, to the widest audience ever. Educational institutions and Companies are exploiting these opportunities, mostly by providing to their students and employees training opportunities via resources, based on fixed contents and fixed learning paths stored on specific portals, called Virtual Learning Environments (VLEs).

Recently a new approach to learning over the Internet has become popular, the so called Personal Learning Environment (PLE). PLE supports students in building their own learning path, freely exploring contents available on the Internet. In this scenario, each learning path is modeled on each student's needs.

PLE represents a customizable and powerful alternative to VLE, but it has a drawback: it requires much more attention about the quality of the information

Luigi Romano
Free University of Bolzano, Italy
e-mail: luigi.romano@unibz.it

P. Vittorini et al. (Eds.): *2nd International Workshop on Evidence-Based TEL*, AISC 218, pp. 101–108.
DOI: 10.1007/978-3-319-00554-6_13   © Springer International Publishing Switzerland 2013

retrieved, and the owner should be able to perform an effective critical analysis of its contents. This is not so easy, for example to students, if they have not been specifically trained. This paper focuses on how to foster student competences in critical analysis of retrieved materials within a PLE, so to improve their 'learning how to learn' skills, in a lifelong learning perspective.

The paper is organized as follows: in section 2 background literature is presented, then in sections 3 and 4 the most relevant aspects of the experience shall be presented, together with gathered data and achieved results. Section 5 will present the final conclusions and possible future developments.

2   Background Literature

The complexity of today's society obliges people to become lifelong learners[7][8], that is, to be able to complement any formal education they have received with non-formal, personalized acquisition of new knowledge and skills. Gaining such new knowledge (i.e. how to complement formal education) is now an indispensable condition for any improvement in professional careers. The new skills include the abilities of working in teams, of collecting information from various sources, and of performing critical analysis on such information.

Internet and the WWW obviously gave a new meaning, and new opportunities, to the concepts of "Lifelong Learning", "working together" and "sharing ideas and projects". By using networked computers, E-Learning communities established all over the world, similarly to the "Communities of Practice" (COP) described by Wenger[17], as "places" where participants continuously improve their knowledge and expertise.

The needed skills of critical analysis and of arguing develop during adolescence, as pointed out by Piaget[11] and Vygotsky[15][16]. In this perspective, Bruner[2] claimed that social interactions and branches of learning are "boosters of culture", and the final purpose of School is to teach students to think in autonomy. "Think in autonomy" is the key competence in a Lifelong learning perspective, since the lifelong student has to build his/her own learning path, and he/she should distinguish among the information reached which elements are useful - and reliable. The technological counterpart of "thinking in autonomy" is a PLE, that is, an agile learning structure, as described by Downes[4], Van Harmelen[14], Attwell[1], Wilson[18] and Chatti[3]. It represents the opportunity for learners to build a personal learning path. If a student has not developed his/her critical sense, Jonassen[5][6] stated that the building of a learning path could be compromised, by adopting false information as true.

Quoting the author "... The conundrum ... for instructional designers, however, is that if each individual is responsible for knowledge construction, how can we as designers determine and insure a common set of outcomes for learning, as we have been taught to do? ...".

3   The Rationale and the Description of the Experience

The question raised by Jonassen, quoted in section 2, provides the rationale for this experience. The goal is to let students experience with PLE creation and use, in order to become aware of educational resource quality. This will foster the skills needed in their future as lifelong learners.

The experience involved a class at the 4th year of a Liceo Scientifico, consisting of 22 teenage students. The topic for the experience was Mathematics, and more in details, Plane Trigonometry. The choice of Mathematics was motivated by the difficulties that students usually face in such a topic, so that they would be encouraged to overcome individual difficulties by a personalized study path. Mathematical skills are indispensable in most careers, as OECD-PISA[10] states that "... citizens routinely face situations in which the use of quantitative or spatial reasoning, or other cognitive mathematical competencies would help clarify, formulate or solve a problem. ...".

The experience needed a preparation phase, which took place during the first semester, and developed during the second semester (February to mid June 2012).

In the preparation phase, Mathematics was taught as traditionally, with lectures and paper-and-pencil exercises. At the same time, students were also trained about: what is a PLE and which are its purposes; how to perform an effective team-working, and how to face Mathematics with an heuristic approach, as described by Polya[12]. During the second semester, students abandoned the traditional way of learning Maths, and undertook the following steps:

1. Each Student built a personal website using GoogleSites (i.e. an own PLE).
2. As homework, students were requested to analyze certain Maths topics, on their own, searching for "sound" web resources.
3. After a few days, students had to share their findings in class, discussing with peers and tutor about the quality and trust of their findings, adopting the heuristic approach
4. Elements considered of good quality, after the discussion, were stored in each student's PLE, in order to build a basic mathematical portfolio.
5. Students' PLEs were completed by each student with additional information, and then, the process restarted on a new topic.

These steps describe a process that builds simultaneously the contents of the PLE and students' skills of critical analysis of information retrieved, with the support of the tutor. The methodological approach for the development of each PLE could be referred to a participatory design approach, because each student (that will be the final user of the PLE) could act for modifying their PLE structure accordingly with their needs.

The acquisition of new knowledge, both in a traditional classroom activity, and via the above approach, is finally assessed in the traditional way, that is, with periodical class tests made during the 1st and 2nd semester.

The metrics for evaluating students' strengthening of "critical sense" is based on the use of Ockham's razor[9], i.e. "pluralitas non est ponenda sine necessitate ponendi" and "entia non sunt multiplicanda praeter necessitatem".

These criteria applied to the materials (links, papers and so on) in order to reduce the overwhelming amount of contents discovered on the Internet (this happened especially in the beginning of the activities). Initially students have been supported by tutor in their analysis, and later in this process they proceeded by themselves. Discarded materials fell basically into the following 4 categories: (a) duplications; (b) materials not relevant to the topic, (c) materials too complex for high school students or (d) materials containing mistakes. All materials with exhaustive contents, easy to be read and to be understood by students have been kept.

Table 1 summarizes the results of Ockham's razor principles, applied to materials collected by students from February to May (During June are held the final examination, by providing traditional tests on mathematical topics):

Table 1 Table 1 Results from Ockham's razor

Month	Amount of Materials per month	Material Kept	Percentage of Kept	Material discarded	Percentage of Discarded
February	93	48	51,61	45	48,39
March	74	40	54,05	34	45,95
April	60	43	71,67	17	28,33
May	55	46	83,64	9	16,36

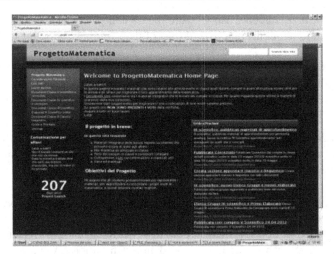

Fig. 1 Screenshot of ProgettoMatematica PLE

To give a feeling of what kind of PLEs have been developed, during step (1) above, the instructor provided students with examples about how to build a PLE, and noticeably with a sample PLE, built by the instructor himself, called "ProgettoMatematica"[13], for a screenshot see Fig.1.

This website contained features like solved exercises, papers and external links. At the end of the schoolyear, all students completed a brief anonymous Questionnaire on their subjective assessment of the experience.

4   Results

The analysis of the results concerns two aspects: the first one was related to the increase of "critical sense" by students and the second one concerned their results in mathematical understanding.

Concerning the increase of "critical sense", from Table 1 we notice that the number of materials found by students on Internet is decreasing from February to May, and at the same time the percentage of materials kept is increasing. This result shows that students are improving their skills of analyzing contents found on Internet, and the effectiveness of the scaffolding provided by the tutor, including the sample PLE. The second result of the experience is given by grades of each student referring to plane trigonometry at the end of the II semester.

The assessment of the mathematical skills of each student in the class was made at the beginning of the 1st semester, around the middle, and at the end of the first semester. It showed a bad situation, with most students having problems, and a resulting low performance in mathematics. Class results are summarized in Table 2.

Table 2 Class situation during the 1st semester

Total amount of students in class	22
Students below par at the end of the 1st semester	12
Students below par with a severe situation	5
Percentage of students below par at the end of the 1st semester	54.50%
Average of marks of the whole class at the end of the 1st semester	5.5

Understanding the motivations behind these poor results required several interviews by the instructor. Most students' problems related to a general lack of methodologies for dealing effectively with mathematical issues. This resulted in poor understanding of proposed Maths topics, in a short term (school success/failure) perspective. What really is worrisome is the consequence of such a deficiency in a long term perspective, i.e. the lack of methodologies to deal with Maths is likely to negatively impact also in their approach to lifelong learning, and in their future careers. Bearing in mind that the objective was well beyond that of simply grading more students as sufficient in Maths, the work described in section 3 was undertaken. As a result, the construction of own PLE caused a change in the whole class attitude w.r.t. the 1st semester. Students' results assessed during the 2nd semester revealed a general improvement in facing mathematical problems.

The reason of this improvement, investigated by further interviews conducted with students, related to benefits caused by the new methodological approach to the mathematical problems proposed.

Results of assessments performed at the end of the 2nd semester are summarized in Table 3.

Table 3 Class situation at the end of the 2nd semester

Total amount of students in class	22
Students below par at the end of the 2nd semester	2
Students below par with a severe situation	2
Percentage of students below par at the end of the 2nd semester	9.10%
Average of marks of the whole class at the end of the 2nd semester	6.13
Increase of the average of marks w.r.t. 1st semester	11.45%

Answers to the questionnaire delivered at the end of the experience have been summarized in Table 4. They show that learners generally enjoyed the experience and especially the development of their own PLE helped them to improve their methodological skills. Emphasizing the importance of acquiring a good methodological approach, OECD[10] refers to skills like "modeling" and "problem-solving" that are clearly related to a well-structured methodological approach to any Mathematical problems.

Table 4 Questionnaire Results

Question	Answers	Percentage
Did you like the new working methodology? [1: No, 2: Not so much, 3: I liked it but there are aspects that could be improved, 4: I liked very much]	Answ. 1: 3 Answ. 2: 0 Answ. 3: 2 Answ. 4: 17	Answ. 1: 13.64% Answ. 2: 0% Answ. 3: 9.08% Answ. 4: 77.18%
How do you think this approach to Mathematics has improved your skills?		
I improved in planning my activities	Answ. 1: 16	Answ. 1: 72.64%
I improved in methodologies for data mining in Internet, concerning my studies	Answ. 2: 1	Answ. 2: 4.54%
It was too much distracting, it did not help me	Answ. 3: 3	Answ. 3: 13.64%
It was indifferent for me	Answ. 4: 2	Answ. 4: 9.08%

5 Conclusion and Future Works

This paper illustrated an experience in designing and implementing a new approach, based on the use of PLE, for fostering the creation of a critical and analytical mind attitude in students, and at the same time improving their mathematical skills.

The results achieved with the approach were encouraging, since traditional assessment of students showed a significant improvement of their mathematical knowledge. Satisfaction of students was also collected via anonymous questionnaire, and showed good reactions to the proposed new approach.

The title of this paper is purposely joking on a very serious issue: that of preparing the young generation for surviving in a quickly evolving world, and giving them tools for improving their careers throughout their entire working life- in the next 40 years, or even more.

The author claims that this approach has fostered students' skills that will be useful in their career paths, since tomorrow's labour market will be dominated by lifelong learners. Even if a careful check of this claim would require a follow-up of the students' entire careers, the improved skills given by acquiring methodologies for facing effectively a mathematical problem would certainly improve their professional chances, whatever job they will choose in the future. The labour market is always eager for more talented workers that can apply critical sense when confronted with new problems, and can systematically search for corresponding solutions, applying heuristics: this is precisely the value added by the described approach, on top of the understanding of plane trigonometry.

Concerning future developments of this approach in the school, it will be important to test it on a broader sample. This approach gradually shifts the focus of study activities from strict tutor control, in an environment of regulated learning patterns, towards the acquisition of competences at a higher level; the latter will be a basic requirement in tomorrow's society, oriented to a lifelong learning perspective.

References

1. Attwell, G.: Personal Learning Environments - the future of eLearning?
   http://www.elearningeuropa.info/files/media/media11561.pdf
   (accessed February 20, 2013)
2. Bruner, J.S.: Toward a theory of instruction. Harvard, Cambridge (1966)
3. Chatti, M.A., Jarke, M., Frosch-Wilke, D.: The future of e-learning: a shift to knowledge networking and social software. Int. Journal of Knowledge and Learning 3, 404–420 (2007)
4. Downes, S.: E-learning 2.0. eLearn Magazine 10 (2005)
5. Jonassen, D.: Thinking Technology. Educational Technology, 34–37 (1994)

6. Jonassen, D.: Designing constructivist learning environments. In: Reigeluth, C.M. (ed.) Instructional Design Theories and Models: A New Paradigm of Instructional Theory, vol. II, pp. 215–239. Lawrence Erlbaum Associates, Mahwah (1999)

7. European Commission Educ. & Training: Lifelong Learning Programme, http://ec.europa.eu/education/index_en.htm (accessed May 20, 2010)

8. Lifelong Learning Programme, http://ec.europa.eu/education/lifelong-learning-programme/doc78_en.htm (accessed May 20, 2010)

9. Ockham, http://www.britannica.com/EBchecked/topic/424706/Ockhams-razor (accessed December 10, 2012)

10. OECD, http://www.oecd.org/pisa/pisaproducts/44455820.pdf (accessed November 15, 2012)

11. Piaget, J.: The Origins of Intelligence in Children. Basic Books, New York (1952)

12. Polya, G.: How To Solve It. Princeton University Press (1945)

13. https://sites.google.com/site/progettomatematica/home (accessed February 19, 2013)

14. Van Harmelen, M.: Personal Learning Environments. In: Proceedings of the 6th International Conference on Advanced Learning Technologies (ICALT), pp. 815–816 (2006)

15. Vygotsky, L.: Mind and Society. Harward University Press, Cambridge (1978)

16. Vygotsky, L.: Thought and Language. MIT Press, Cambridge (1962)

17. Wenger, E.: Communities of practice: learning, meaning, and identity. Cambridge University Press (1999)

18. Wilson, L., Johnson, M., Beauvoir, P., Sharples, P., Milligan, C.: Personal Learning Environments: Challenging the dominant design of educational systems. Journal of e-Learning and Knowledge Society 3(2), 27–38 (2007)

Test My Code: An Automatic Assessment Service for the Extreme Apprenticeship Method

Arto Vihavainen, Matti Luukkainen, and Martin Pärtel

Abstract. We describe an automated assessment system called Test My Code (TMC) which is designed to support instructors' and students' work in programming courses that use the Extreme Apprenticeship method. From the students' point of view TMC is a transparent assessment service that is integrated to a industry-standard programming environment. TMC allows scaffolding during students' working process, retrieves and updates exercises on the fly, and causes no overhead to the programming process. From the instructors' perspective, TMC allows collaborative crafting of exercises with small goals that combine into bigger programs, gathering snapshots from students' actual programming process, and collecting feedback. TMC has been successfully used in MOOCs in programming as well as in our university courses.

1   Introduction

Methods based on Cognitive Apprenticeship (CA) [8] have recently had success in teaching programming [2, 7]. Extreme Apprenticeship (XA) is an interpretation of CA that builds a relationship between the teacher and the learner, and emphasizes scaffolding by providing support from material, exercises and people around the student [17]. A key success factor in XA is the students' own activity and the amount of deliberate work put in working on the exercises. In a XA-based course there are usually lots of small exercises that allow the student to experience success, build her self-confidence, and hone her programming routine. Small exercises combine into bigger entities, explicitly showing the working process needed to solve larger problems. After a student has worked through a problem in steps, she can later solve similar problems without the step-wise scaffolding.

Arto Vihavainen · Matti Luukkainen · Martin Pärtel
University of Helsinki, Department of Computer Science
P.O. Box 68 (Gustaf Hällströmin katu 2b)
FI-00014 Helsinki
e-mail: {avihavai,mluukkai,partel}@cs.helsinki.fi

P. Vittorini et al. (Eds.): *2nd International Workshop on Evidence-Based TEL*, AISC 218, pp. 109–116.
DOI: 10.1007/978-3-319-00554-6_14   © Springer International Publishing Switzerland 2013

It is important that any work done is as "genuine" as possible as one should not have to later unlearn learned skills and practices [4, 10]. In our context learning to program starts with setting up an industry-strength programming environment that professionals use, which emphasizes that the learners are on a path to become true practitioners starting from day one.

Assessing the students' work when each student can work on over 30 programming exercises during a single week is laborious. Although most of the effort can be efficiently spent [14], we have noticed that a noticeable amount of time goes into trivial scaffolding. In early XA courses, the practice was that when a student felt that her exercise was done, it was manually checked by a course instructor and marked down in a course logbook [17].

It is evident that XA can benefit from the use of automated tools. In order to focus the advisors' time on purposeful scaffolding, a minimally intrusive assessment system that supports XA workflow is essential. Our main requirements for the assessment system were that the system (1) should not introduce any additional overhead to the students' working process, (2) should be integrated to an industry-standard programming environment, (3) must allow building of scaffolding into an exercise, (4) must allow awarding points from completing smaller goals, not just from a complete exercise, (5) should cause no additional overhead to course instructors from the management perspective, and (6) should allow honing of software engineering practices for the course personnel as well.

The solution that we describe in this article removes the need for trivial exercise checking, provides basic scaffolding capabilities, and allows the instructors to focus more on their actual work – mentoring and supporting the students. The system is integrated to an IDE, and allows dissecting the students' working process even if they cannot be present. In order to verify that TMC can be used in long-distance education, we have used it in our introductory programming (CS1) courses as well as in MOOCs in programming [16].

This document is organized as follows: Section 2 gives a more detailed explanation of XA in the context of programming courses. Section 3 describes the Test My Code-server from both students' and instructors' perspective, and section 4 gives an explanation on how exercises can be crafted for TMC. Section 5 gives an overview of our initial evaluations, and in the final section we describe future work that is planned to further improve TMC.

2   Extreme Apprenticeship Method

The Extreme Apprenticeship (XA) method extends Cognitive Apprenticeship [8] by bringing in practices from methods such as Extreme Programming (XP) [3], and emphasizes students' personal effort and bi-directional communication between the learner and the advisor. Core values in XA are [17]: (1) A craft can only be mastered by actually practicing it, for as long as is necessary, and (2) continuous feedback between the learner and the advisor. The advisor must be aware of the successes and challenges of the learner throughout the course, which allows the advisor to provide better scaffolding and further improve the course material.

Instruction is usually focused to labs where the students are scaffolded as they work on course-related exercises. By scaffolding we mean that the students are supported in a way that they are not given direct answers, rather, just pushed into a direction to discover the answers themselves. The course instructors focus on providing high-quality learning material, exercises, and enough guidance to help students during their working process. The course material, for example, emphasizes the working process using worked examples [7] and process recordings [5], both explicitly describing *how* a program is crafted using stepwise sub-task division: one must always start small to grow big. The scaffolding that is built into the exercises and material makes it easier for the student to proceed *and* to a favorable direction: learning achievements are made visible to the student, which further increases the motivation to continue [6].

As a craft can only be mastered by actually practising it, our semester-length programming course has over 350 programming tasks, ranging from simple "Hello World's" to programming a snake game from scratch. A student must complete the majority of the exercises in order to receive a good grade. In order to verify that a student has worked on the exercises herself, we organize either a written or an oral exam at the end of the course.

Most of the exercises are composed of small incremental tasks that combine into bigger programs. Incremental tasks are used to imitate a typical problem solving process: students explicitly practise programming but are constantly influenced by the written-out thought process behind the pre-performed sub-task division. Exercises are intentionally written out to be as informative as possible, and often contain sample input/output descriptions and code snippets with expected outputs that provide further support for verifying the correctness of the program. This also helps students learn to read code written by others.

Scaffolded exercises bring several benefits for the students. Getting started is easier since exercises are split into smaller tasks. By working her way through several scaffolded exercises each week, the student starts to see how programs should be structured in addition to practicing fundamental programming routines. Exercises are also used to direct students away from bad programming habits such as the use of unnecessary class attributes and unclear method names.

Since the ultimate goal of university level instruction should be that the students can master problem solving by themselves, fading is important. Fading means correctly timed gradual dismantling of scaffolds [8], and it is realized in our programming courses via open ended exercises that do not hint or enforce any kind of program structure, but only define how the application is supposed to work for a given user input, e.g. by defining a UI in a relatively strict manner. Before venturing into open ended exercises, the students have worked through similar problems with scaffolding.

3 Test My Code

The first documented automatic assessment systems are from the 1960s, after which a large body of assessment services have been created, see e.g. [9, 1, 12]. Several of the existing systems did fulfill some of our requirements, perhaps the closest match being Marmoset [15]. Despite testing several candidates, we did not find a system that would suit our specific purposes. Having just an assessment service would waste tons of experience that we have gathered while scaffolding students: surely some of the experience could be built into a system.

Work on Test My Code (TMC) started in 2011 as two bachelor-level capstone-projects, where one team was assigned to work on an assessment server, and one team on an IDE plugin. During the projects, the teams built a working prototype, which has since been further developed into a flexible assessment service that integrates seamlessly into the programmers' workflow. TMC can be deployed to cloud environments, and it supports the creation of exercises with built-in scaffolding.

TMC is based on client-server architecture, see Figure 1. On the client side the user has a NetBeans plugin that: (1) retrieves and updates the exercises from a server, (2) provides scaffolding and runs tests locally, (3) submits exercises to the server, and (4) gathers snapshots from the students' working process. Snapshot gathering can be enabled or disabled by the student. In addition to the plugin, TMC has a web interface that students use for creating and administering their user account as well as viewing suggested solutions. Course instructors use the web interface for administrative tasks and viewing statistics and feedback. The web interface can also be used for submitting exercises.

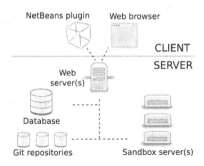

Fig. 1 High-level architecture of TMC

On the server side, the system is split into several components. On the front there exists one or more web servers that are connected to a database, one or more version control repositories (in our case Git) that contain the exercises, and a set of sandbox servers. The database stores information of courses, exercises, users, submissions, scores and snapshots. Each course is located in a separate version control repository, where the exercises are added and updated by the course instructors. The sandbox servers receive submitted exercises from the web server, run them on sandbox instances (transient user-mode linux virtual machines), and submit the results back to the web server. In order to handle high-demand situations (e.g. deadlines), TMC has a queuing mechanism, and increasing the amount of sandbox instances is easy.

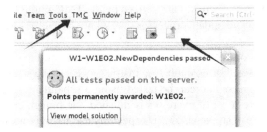

Fig. 2 New menu option and buttons are installed by the TMC-plugin. The smiley-window is an indicator that previously submitted exercises passed the tests.

Student's perspective

From the students' perspective, working starts with registering to a course using the web interface. Once the registration is done, basic development tools, e.g. NetBeans, are installed, after which the TMC-plugin is added to the IDE. This is usually the very first exercise for a course that utilizes TMC.

After the TMC plugin has been added to NetBeans, a new menu option entitled TMC and three new toolbar buttons are visible in the IDE (see areas pointed out with arrows in Figure 2). The menu option gives possibilities for changing settings (e.g. username, course, directory for downloading exercises), checking for new exercises, and submitting answers. The leftmost toolbar button is for running the current application (independently of any accidentally selected main project). The middle button is for testing the application locally, and the third button is for sending the solution to the assessment server. If the student presses the run tests locally-button, local tests for the exercise are run and possible scaffolding messages are shown.

Since one of the goals in our CS1 is to introduce students to the concept of unit testing, the scaffolding messages follow the classic TDD style [11], where the student sees a green/red bar and an informative error message. TMC supports both local and server-side tests, which means that the test code can be hidden from the student. Once an exercise has been submitted, the student will see feedback from the assessment server after related tests have been executed, for example the "All tests passed"-notifier, which is visible in the Figure 2. One can configure feedback questions to the notifier, which allows easy feedback gathering.

Instructor's perspective

From the instructors' perspective creating a new course starts with initiating a Git version control system repository that is used to store the exercises. Once a repository has been created, a new course instance is created using the TMC web-interface, that provides options for e.g. configuring feedback questions shown to students as they submit exercises, inspecting students' code and viewing score and submission statistics.

Each instructor wishing to contribute to the exercises clones the git repository to a local machine. Once modifications are made or new exercises have been crafted, changes are pushed to the central repository. An administrator then asks TMC to load the changes from the central repository and makes them visible to the students.

Any changes to existing exercises are noticed by the TMC-plugin, and offered to the students as updates. The update process is careful not to overwrite any code written by the student.

Using a version control system increases productivity in the creation and publication of the exercises. For example, the instructors can have a separate branch for the exercises-in-progress that can be tested using a separate course for instructors and volunteers. Once a set of exercises is finished, the branch can be merged to a branch that is visible to students. This is especially useful for educators that perform team-teaching.

The TMC web interface provides online statistics on students' progress, as well as an easy access to student feedback. Instructors can respond to feedback with a single click. The feedback gives valuable insights into the suitability of exercises, e.g., is the amount of scaffolding optimal, are the instructions unambiguous, are the scaffolding tests giving understandable diagnostics. Aggregate statistics are available as well, e.g. overall student activity over time.

4   Crafting TMC Tests

Crafting an exercise starts with the creation of a NetBeans Java project, and pushing the project folder into the course git repository. Each exercise is iteratively developed using TDD and JUnit, simulating the programming process that a student should perform. Tests are created for each increment in the exercise, and used to validate the programming process. When the model solution to the exercise is created using TDD, the initial tests are done as a by-product.

As the project that is used as a model solution will be the same project that the student receives as a starting point, additional modifications are needed. Source files or specific content in files are removed by the TMC server based on annotations in the code, and providing stub code is also supported.

As quite a few of the basic validations are alike, TMC includes a testing specific domain specific language (DSL) library that provides convenient access to students' code via reflection. It also raises user-friendly errors in some common scaffolding situations, e.g. "did you create the class MainProgram in the correct package?". To encourage good code quality, inner structure of a class is also tested. For example, using more fields than needed is a common novice mistake, and students can be easily scaffolded away from it.

Overall, the tests are built in a way that gives direct support to the incremental nature of the exercises. They help students to focus on progressing in small steps even within a single exercise: it is important to concentrate on making tests pass one by one in a meaningful order.

5   Evaluation

We have evaluated TMC in our courses as well as in MOOCs in programming [16]. The number of students who have used TMC so far exceeds 2000, and TMC has been used to evaluate over 125 000 submissions.

We have evaluated students' satisfaction with TMC using anonymous feedback from our courses. When the first beta of TMC was released in September 2011 for a 200 student CS1 course, one half of the course exercises were assessed using TMC. Only 58% of the feedback concerning the use of TMC was positive. Based on the experiences and feedback from the first course, major improvements were made. Since finishing the improvements in January 2012, the student feedback regarding TMC has been very good: 80% of the feedback regarding TMC during spring 2012 CS1 was highly positive, and in our first MOOC in programming it even got better, most of the feedback was praising and only 9% had some complaints. The majority of the negative comments were not severe, usually complaining about minor details such as "it was quite irritating when TMC demanded a questionmark..".

During fall 2012, 79% of the comments regarding to TMC were positive or highly positive. The subtly negative comments were from early parts of the course, which leads us to assume that some of the feedback can be explained by the students struggling with learning a programming language, and not understanding the scaffolding messages properly. In the latter parts of the course (final 6 weeks), feedback related to TMC has only been positive. We have also observed (in person, through IRC channels and emails) several spontaneous testimonials for the superiority of TMC when compared to current assessment automata used in several other universities and certain MOOCs.

When considering the educational value of TMC, we must look at the impact of XA in our programming courses. Before XA was introduced, the pass-rate average for our CS1 course has been 55.49% over 16 course instances. After introducing XA, average pass-rates before TMC have been 73.45% over 3 course instances, and after applying TMC in our XA courses the average pass-rates are currently at 75.76% when averaged over 2 course instances.

6 Conclusions and Future Work

We have described TMC, an automatic assessment system that seamlessly supports XA-style programming courses where the emphasis is on meaningful, scaffolded exercises and bi-directional communication between students and instructors. We believe that the main success factors of TMC lie in (1) the multi-level feedback mechanism of XA, (2) the use of industry-level programming environment, (3) scaffolding provided by the tests, and (4) the small goals inside the bigger goals.

TMC has improved our course instructors work by making it more meaningful. By removing trivial exercise checking and scaffolding, instructors can spend more time on more demanding scaffolding tasks. TMC helps advisors by providing meaningful output that can be used for scaffolding students. Using TMC has also enabled us to organize pedagogigally meaningful MOOCs in programming.

Using an automated assessment system has also its negative sides. We have observed that some of our students rely too much on automatic scaffolding and do not write spontaneous test-programs of their own. Due to the XA context this is not as problematic as it could be – the advisors that monitor the students' progress scaffold them to create test-programs, pushing the students to think outside of their own box.

Guiding students towards better design has been so far done using Java's reflection and by instructors. We are considering integrating static analysis tools such as Checkstyle into TMC, which would make addressing code conventions (e.g. indentation, variable naming, method length and complexity) easier. In addition, we are currently developing a deterministic profiler similar to [13] which makes it possible to conduct repeatable benchmarks of students' algorithms.

References

1. Ala-Mutka, K.: A survey of automated assessment approaches for programming assignments. Somputer Science Education 15(2), 83–102 (2005)
2. Astrachan, O., Reed, D.: AAA and CS 1: The applied apprenticeship approach to CS 1. SIGCSE Bulletin 27 (1995)
3. Beck, K., Andres, C.: Extreme Programming Explained: Embrace Change, 2nd edn. Addison-Wesley Professional (2004)
4. Begel, A., Simon, B.: Struggles of new college graduates in their first software development job. In: Proc. of the SIGCSE 2008. ACM (2008)
5. Bennedsen, J., Caspersen, M.E.: Exposing the programming process. In: Bennedsen, J., Caspersen, M.E., Kölling, M. (eds.) Reflections on the Teaching of Programming. LNCS, vol. 4821, pp. 6–16. Springer, Heidelberg (2008)
6. Bergin, S., Reilly, R.: The influence of motivation and comfort-level on learning to program. In: Proc. of the PPIG 2005 (2005)
7. Caspersen, M.E., Bennedsen, J.: Instructional design of a programming course: a learning theoretic approach. In: Proc. of the ICER 2007. ACM (2007)
8. Collins, A., Brown, J.S., Holum, A.: Cognitive apprenticeship: making thinking visible. American Educator 6 (1991)
9. Douce, C., Livingstone, D., Orwell, J.: Automatic test-based assessment of programming: A review. J. Educ. Resour. Comput. 5(3) (2005)
10. Fox, A., Patterson, D.: Crossing the software education chasm. Commn. ACM 55(5), 44–49 (2012)
11. Gamma, E., Beck, K.: Junit: A cook's tour. Java Report 5(4), 27–38 (1999)
12. Ihantola, P., Ahoniemi, T., Karavirta, V., Seppälä, O.: Review of recent systems for automatic assessment of programming assignments. In: Proc. of the 10th Koli Calling. ACM (2010)
13. Kuperberg, M., Krogmann, M., Reussner, R.: ByCounter: Portable Runtime Counting of Bytecode Instructions and Method Invocations. In: Proc. of the 3rd International Workshop on Bytecode Semantics, Verification, Analysis and Transformation, ETAPS 2008 (2008)
14. Kurhila, J., Vihavainen, A.: Management, structures and tools to scale up personal advising in large programming courses. In: Proc. of the SIGITE 2011. ACM (2011)
15. Spacco, J., Hovemeyer, D., Pugh, W., Emad, F., Hollingsworth, J.K., Padua-Perez, N.: Experiences with marmoset: designing and using an advanced submission and testing system for programming courses. In: Proc. of the ITICSE 2006. ACM (2006)
16. Vihavainen, A., Kurhila, J., Luukkainen, M.: Multi-faceted support for mooc in programming. In: Proc. of the SIGITE 2012. ACM (2012)
17. Vihavainen, A., Paksula, M., Luukkainen, M.: Extreme apprenticeship method in teaching programming for beginners. In: Proc. of the SIGCSE 2011. ACM (2011)

Author Index